RIVER AND I
OF THE GR...

Kristin Huisinga
Lori Makarick
Kate Watters

With a Foreword by Ann Zwinger

2006
Mountain Press Publishing Company
Missoula, Montana

© 2006 by Kristin Huisinga, Lori Makarick, and Kate Watters

First Printing, September 2006
All rights reserved

Foreword © 2006 by Ann Zwinger
All photographs and illustrations credited where they appear.

Front cover photo © 2006 by Glenn Rink: Newberry's Yucca *Hesperoyucca newberryi*
Front cover design by Raechel Running, RMRFotoarts.com

Back cover insets, top to bottom:
Sacred Datura *Datura wrightii* —Photograph © 2006 by Bob Rink
Texas Storksbill *Erodium texanum* —Photograph © 2006 by Kathy Darrow
Mariposa Lily *Calochortus flexuosus* —Photograph © 2006 by David Edwards

Back cover cutout:
Grand Canyon Beavertail Cactus *Opuntia basilaris* var. *longiareolata* —Photograph © 2006 by Kyle George

Library of Congress Cataloging-in-Publication Data

Huisinga, Kristin, 1974–
    River and desert plants of the Grand Canyon / Kristin Huisinga, Lori Makarick, Kate Watters ; with a foreword by Ann Zwinger.
    p. cm.
    Includes bibliographical references and index.
    ISBN-10: 0-87842-523-3   ISBN-13: 978-0-87842-523-5 (pbk. : alk. paper)
    1. Stream plants—Arizona—Grand Canyon—Identification. 2. Desert plants—Arizona—Grand Canyon—Identification. 3. Grand Canyon (Ariz.)   I. Makarick, Lori, 1969– II. Watters, Kate, 1971–   III. Title.
QK147.H85 2006
581.9791'32—dc22

2006021922

PRINTED IN HONG KONG BY MANTEC PRODUCTION COMPANY

Mountain Press Publishing Company
P.O. Box 2399 • Missoula, Montana 59806
(406) 728-1900

In these troubled times when all the world is so upset, I think it rather steadies one to 'consider the lilies of the field,' to examine closely the exquisite texture and coloring of the petals of many flowers, to look with appreciation upon a magnificent forest and to feel thankful for the beautiful world in which we live.

—ROSE COLLOM, 1870–1956

This book is dedicated to those who have come before us, to the community and fellowship of the present, and to the future caretakers of this place of pure marvel, the Grand Canyon.

We pay homage to Rose Collom, Grand Canyon National Park's first botanist; Elzada Clover and Lois Jotter Cutter, first botanists to formally collect and describe plants along the Colorado River; and Wendy Hodgson, who continues the legacy of all three with enough passion to inspire us all.

Lorin Bell, Elzada Clover, and Lois Jotter Cutter collecting on their river trip in 1938 —Lois Jotter Cutter creator, NAU. PH.95.3.31, Cline Library Special Collections & Archives

Rose Collom collecting on the North Rim in the early 1940s —Grand Canyon National Park Museum Collection 52846

Wendy Hodgson collecting in the Grand Canyon
—Photograph © 2006 Kate Watters

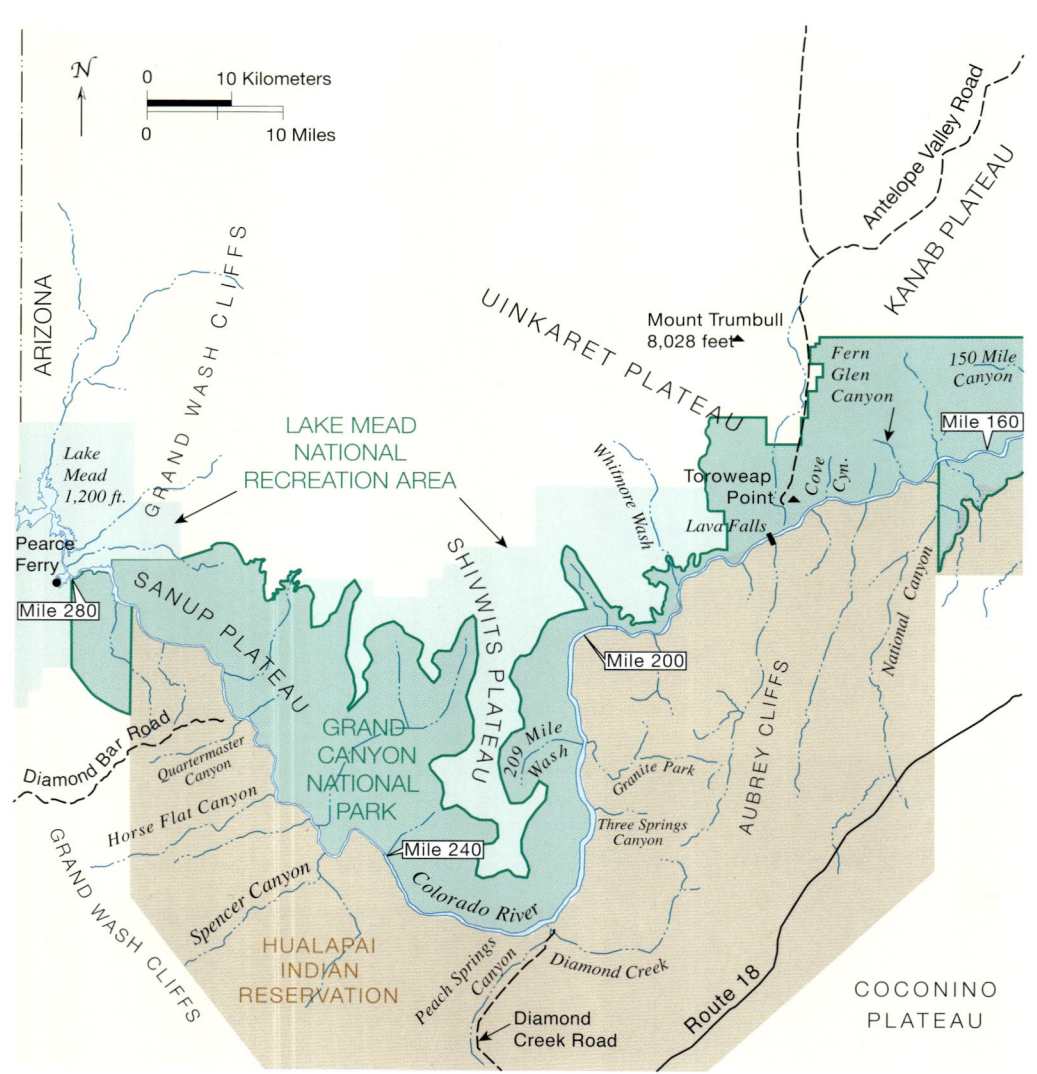

Grand Canyon —Map by Mountain Press Publishing Company

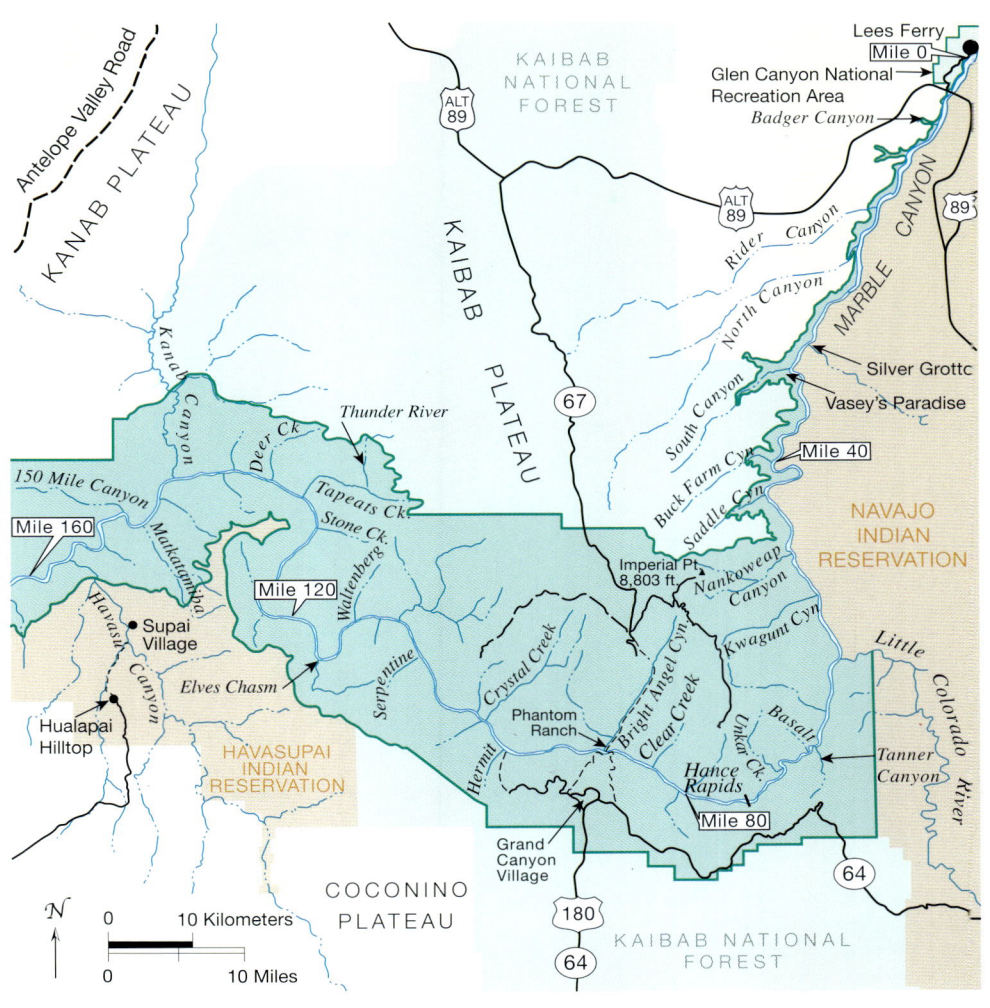

Utah Century Plant *Agave utahensis* —Watercolor © 2006 Samuel P. Jones

CONTENTS

MAP OF THE GRAND CANYON *iv*
FOREWORD *xi*
ACKNOWLEDGMENTS *xiii*
CONTRIBUTORS AND SUPPORTERS *xiv*
HOW TO USE THIS BOOK *1*
Thumbnail Identification Guide *2*
GRAND CANYON ECOLOGY *9*
Grand Canyon Plant Communities *9*
Riparian Vegetation Then and Now *13*
Desert Plant Adaptations *14*
Growing History: People and Plants in the Grand Canyon *16*
The Foundation of Life: Algae and Macrophytes of the Colorado River *18*
Biological Soil Crust *19*
Hot Spots of Diversity: Springs in the Grand Canyon *20*
Seeking the Sweet Reward: A Pollinator Vignette *21*
Biodiversity in Peril: Exotic Species in the Grand Canyon *23*
Treasures of Diversity: Rare Plants in the Grand Canyon *25*

FERNS AND FERN ALLIES *26*
Horsetail Family *26*
Maidenhair Family *27*

GRASSES AND GRASSLIKE PLANTS *30*
Cattail Family *30*
Grass Family *32*
Rush Family *52*
Sedge Family *54*

TREES *57*
Bignonia Family *57*
Birch Family *58*
Bittersweet Family *58*
Buckthorn Family *60*
Citrus Family *60*
Cypress Family *62*
Elm Family *64*
Maple Family *64*
Oak Family *66*
Olive Family *66*

Pea Family 68
Tamarisk Family 74
Willow Family 74

SHRUBS AND FORBS 78
Agave Family 78
Amaranth Family 84
Barberry Family 84
Bellflower Family 86
Bittersweet Family 88
Borage Family 88
Broomrape Family 92
Buckthorn Family 92
Buckwheat Family 94
Buttercup Family 96
Cactus Family 98
Caltrop Family 110
Carrot Family 110
Citrus Family 112
Dogbane Family 112
Ephedra Family 114
Evening-Primrose Family 116
Four O'Clock Family 118
Gentian Family 122
Geranium Family 122
Goosefoot Family 124
Grape Family 128
Lily Family 128
Madder Family 130
Mallow Family 132
Malpighia Family 132
Milkweed Family 134
Mint Family 136
Mustard Family 138
Nightshade Family 144
Nolina Family 150
Ocotillo Family 152
Oleaster Family 154
Orchid Family 154
Pea Family 156
Phlox Family 162
Plantain Family 166
Poppy Family 168
Primrose Family 168
Ratany Family 170
Rose Family 170
Snapdragon Family 174

Spurge Family   *180*
Stickleaf Family   *182*
Sumac Family   *184*
Sunflower Family   *186*
Vervain Family   *224*
Waterleaf Family   *226*

PLANT ANATOMY ILLUSTRATIONS   *229*

GLOSSARY   *235*

SELECTED BIBLIOGRAPHY   *243*

INDEX   *249*

The desert in bloom —Photograph © 2006 by Bert Jones

Springs habitat in a Elves Chasm
—Photograph © 2006 by Bronze Black

FOREWORD

As long as I've traveled the Colorado River, I've happily answered questions such as, "What's that big white trumpet-shaped flower that opens early in the evening?" and "Those look like orchids, but down here?" or "What kind of a cactus is *that*?" At long last here's a book that answers all those questions and more in a thorough, attractive, and easy-to-use field guide that covers those plants that river passengers and hikers see.

A book of such quality and breadth could only be written by three botanists, with the input of more than eighty collaborators who range from botany professors, professional writers, restoration ecologists, consultants, self-taught plant lovers, artists, photographers, illustrators, graphic designers, boatmen, and naturalists—an amazing array of contributors. All have spent time in and researched the canyon carefully and have an experienced eye for what a reader needs to identify a plant. Identification is made easier as both photographs and drawings illustrate plants. Nor have the authors neglected that army of grass-like plants for whom "sedges have edges and rushes are round, and grasses are hollow from their knees to the ground" was coined.

Rounding out the portrait of each plant is a description of its preferred habitat, the places you're most likely to see it. And then they add the kind of fascinating, well-researched tidbits that their readers will relish.

I'm all for learning the scientific names, first and last, just as we know our friends' names, first and last. Oftentimes the generic or specific name tells something about the plant, either an obvious characteristic or a person's name that carries a little bit of history with it. Knowing the scientific names avoids the confusion of common names, because there is only one plant with the precise name. In this book the reader will find that *Castilleja* is the generic name of Indian paintbrush, named after a Spanish botanist. Scratchgrass (*Muhlenbergia asperifolia*) was named after botanist G. H. E. Muhlenberg, and the specific name refers to the roughness of the leaves, a lesson in a name.

Despite the fact that all this information is enticing and intriguing, the main thing that attracts us to plants in the first place is their beauty. Cup your hands around a furled evening-primrose at dusk and feel the flutter of its opening petals against your fingers as it responds to the warmth of your hand, then inhale the delicate fragrance. Observe monkey flowers with their companions, columbines and maidenhair ferns, glisten and sparkle by a seep as drops of water fall on their blossoms. Watch a hawk moth center in on a datura flower freshly unfurled at dusk. Check out the small flies that are immobilized on a spiny aster when the sun goes behind a cloud. Rub a mint leaf between your fingers, or take a sniff of poreweed and get a noseful of an unforgettable aroma.

Cultivate an acquaintance with a sand verbena languishing in the warm sand and let its perfume sweeten the day. Examine its sand-laden leaves and the small, individual tube flowers—make its acquaintance. Spend some time with it, speak its name, *Abronia*, and let its four syllables march off your tongue. Greet it each time you see it, thank it for what it adds to your life. Enjoy being in touch with the world around you. Believe in this beauty and reassurance that flowers below the rim carry, take time to touch a furry leaf of brittlebush, taste a healing oil, listen to the breeze rustle through a cottonwood's leaves. Consider yourself blessed to be here, one of the privileged who understand what William Blake, writing two hundred years ago knew:

> To see a World in a Grain of Sand
> And a Heaven in a Wild Flower,
> To hold Infinity in the palm of your hand,
> And Eternity in an hour.

—ANN ZWINGER

Sacred Datura (*Datura Wrightii*) unfurling as the day cools
—Photograph © 2006 by John Running

ACKNOWLEDGMENTS

We are indebted to our technical reviewers and mentors: Tina Ayers, Nancy Brian, Dan Hall, David Hammond, Wendy Hodgson, Rose Houk, Art Phillips, Glenn Rink, Randy Scott, John Spence, and Ann Zwinger. You volunteered countless hours marking up pages on our many drafts, pushing us to become better writers. We checked our facts a little more closely after seeing your red marks and reading your sassy comments. "No!" "Really?" "I've never seen this." "Seems to or does?" "What meteor collision?" You added intricacies and fascinating details about each plant, helping us bring them to life. You also encouraged us to use the active voice, which makes the book easier to read.

The voices of many authors inspired us to write creatively about plants and to find interesting ways to tell their stories: Sharman Apt Russell, Janice Bowers, the many contributors to *A Natural History of the Sonoran Desert*, and Ann Zwinger.

We are grateful to our funders who steadfastly supported this project from the beginning and without whom this endeavor would not have been possible. Grand Canyon Conservation Fund, Arizona Native Plant Society, Flagstaff Cultural Partners, and T&E Inc. awarded generous grants to this venture. Special thanks to Lynn Hamilton at Grand Canyon River Guides, Inc. for serving as our fiscal agent.

Jennifer Carey, our editor at Mountain Press Publishing Company, escorted us through the four-year process of bringing a book into the world with patience and encouragement.

Without Michael L. Charters' website "California Plant Names: Latin and Greek Meanings and Derivations" you would never know interesting tidbits about the naughty nymphs and Veronica's handkerchief.

Special thanks to Ronald Taylor, "a stickler for botanical accuracy" and author of many of our favorite field guides, who came to us post-humously via Jennifer to help with the final task of compiling our glossary.

All the chocolate in the world would not be enough to thank Colleen Hyde, who tirelessly provided us with electronic spreadsheets of all the plant collections housed in Grand Canyon National Park's Museum Collection.

David Edwards, Geoff Gourley, and Bill Hatcher provided early guidance that helped us capture and choose better photographs and emphasized the importance of consistency and quality of the images in this book.

Joshua Lorentz and Susan Zaunbrecher shared their understanding of the law, helping us comprehend all of the legal jargon of the book contract. We thank John Rimel at Mountain Press Publishing Company for being flexible with the contract, allowing us to maintain our commitment to the contributors who offered their creative vision through words and images.

Throughout the writing process we found it necessary to sequester ourselves in faraway places where phones didn't ring and everything we needed in life fit into six colorful woven bags. Special thanks to Diane Ross and Don and Donna Huisinga for opening your homes for our book retreats.

We are especially thankful for the rain clouds that blessed the landscape in 2005, urging forth spectacular blooms. Without them, this book would be conspicuously less colorful.

CONTRIBUTORS AND SUPPORTERS

This project is long overdue and has been conceived many times by some of the contributors credited here and others who are passionate about Grand Canyon plants. Many have had visions of what this book should look like and we tried to incorporate as many suggestions as possible into this publication. The stories we are passing on were harvested, in pieces, from other books, local floras, regional floras, national floras, herbarium collections, essays, scientific papers, and herbal books. But perhaps more importantly, we garnered juicy tidbits from late-night river conversations and rumors on the street, herbalists, boatmen, librarians, botanists, naturalists, artists, high school kids, teachers, graduate students, former and present staff from Grand Canyon National Park, friends, husbands, family, and people who are fanatical in their love for plants.

The human personalities and experiences brought to the research and production of this book are as diverse as the plant community to which it is devoted. This project represents a cooperative effort by more than eighty-five different individuals: photographers, artists, botanists, river guides, ecologists, archaeologists, and naturalists.

Some unnamed people did not formally contribute but supported us by stopping us in the grocery store or along the trail to ask, "how's that book coming along?" These simple words maintained our momentum and repeatedly confirmed the need for this book.

Due to the collaborative nature of this project, a portion of the proceeds from the sale of this book will go to nonprofit organizations dedicated to education and protection of the Grand Canyon.

Mary Allen	Bob Dye	David Inouye
Tina Ayers	David Edwards	Roeland Jansen
Matthew Berry	Rob Elliott	Ted Johnson
Emma Benenati	Helen Fairley	Bert Jones
Joseph Bennion	Kim Fawcett	Samuel P. Jones
Bronze Black	Claire Fuller	Lisa Kearsley
Dean Blinn	Kyle George	Mike Kearsley
BJ Boyle	Jocelyn Gibbon	Emily King
Nancy Brian	Geoff Gourley	Kristin Kolanoski
Debra Burns	Lisa A. Hahn	Max Licher
Kyle Christie	Dan Hall	Mimi Murov
Lora Colten	David Hammond	Lynn Myers
Regan Dale	Bill Hatcher	Gary Paul Nabhan
Roger Dale	Charly Heavenrich	Arthur M. Phillips III
Kathy Darrow	Mar-Elise Hill	Fred Phillips
Rebecca DeGroot	Wendy Hodgson	Joe Pollock
John Dewine	Marisa Howe	Meredith Potts

Richard Quartaroli
Barb Ralston
John Randall
Suzanne Rhodes
Barry Rice
Bob Rink
Glenn Rink
Daniela Roth
John Running
Raechel Running
Al Schneider

Susan Schroeder
Randy Scott
Jean Searle
Joseph Shannon
Ellie Soller
Celia Southwick
John Spence
Cameron Staveley
Laurie Lee Staveley
Lawrence E. Stevens
Denis Stratford

Harlan Taney
Therean Taylor
Alexandra Thevenin
Kate Thompson
Steve Till
Pamela Walls
Tyler Williams
Michael Yeatts
Sheila Yokers
Ann Zwinger

Utah Century Plant *Agave Utahensis* —Photograph © 2006 by David Edwards

GRASS
Phragmites
Phragmites australis
—Photograph © 2006 by Geoff Gourley

TREE
Fremont Cottonwood
Populus fremontii
—Photograph © 2006 by David Edwards

FERN: **Maidenhair Fern** *Adiantum capillus-veneri* —Photograph © 2006 by Geoff Gourley

FORB: **Mojave Aster** *Xylorhiza tortifolia* —Photograph © 2006 by David Edwards

HOW TO USE THIS BOOK

Plants in this book are organized into four distinct lifeform groups: ferns and fern allies, grasses and grasslike plants, trees, and shrubs and forbs. **Ferns and fern allies** do not have flowers. Ferns reproduce by spores, which are housed in clusters hidden on the undersides of the leaves. Fern allies, which include horsetails, bear spores in cone-shaped spikes atop stems or in leaflike organs in between their leaves. **Grasses and grasslike plants** include grasses, sedges, rushes, and cattails. The herbaceous plants have parallel-veined leaves and generally inconspicuous flowers, often enclosed from below by simple bracts. **Trees** are woody, usually tall plants, often having trunks and branched crowns. A few of the trees described in this book are shrub size, but we included them in the tree section because they are in the same genus as other tree species. For instance coyote willow (*Salix exigua*) is quite shrubby, but is discussed with Goodding's willow (*Salix gooddingii*), a tree whose shade you would appreciate on a hot desert day. **Shrubs** are woody plants that branch at or near the ground level and are smaller than trees. **Forbs** generally do not have woody stems. Shrubs and forbs are grouped, however, because some forbs can have a "shrubby" habit and may be stiff or even woody at the base.

Within each lifeform, we grouped plants by common family name. A quick reference key on the inside front and back covers will help you find the common family name if you know the scientific family name. Color thumbnails at the end of this section help readers find plants by flower color. If there has been a recent change of family name, the old family name is given second. For example, the pea family changed from Leguminosae to Fabaceae in recent years and is listed as Fabaceae/Leguminosae. We followed the accepted family names from *Flora of North America*.

Within each family, plants are organized alphabetically by their scientific name. Scientific names have various parts, including the genus name and the specific epithet with some having varieties (var.) or subspecies (ssp.) For Utah century plant, the genus name is *Agave*, the specific epithet is *utahensis*, and there are two varieties, var. *utahensis* and var. *kaibabensis*. Plant taxonomists are always learning more about how plants are related, which results in a never-ending process of reorganizing and sometimes renaming species according to their physical appearance and genetic makeup. We followed the accepted taxonomy provided on the Integrated Taxonomic Information System, an online database. When our local and regional floras listed synonyms for plants, we included them in parentheses below the accepted scientific name. We did not list all synonyms.

Most plants have more than one common name and we used the ones that are most familiar in the Grand Canyon region. Common names create confusion because many plants bear the same common name. When in doubt or using other field guides, check the scientific name to confirm the species identity.

Each plant description includes a technical section for those plant enthusiasts who bend down low with their hand lenses and rulers. The section includes information about the plant form, leaves, flowers, fruit, flowering season, and elevation range. All botanical measurements are given in metric units (millimeters, centimeters, meters), following the format of most standard floras. A ruler is printed on the inside cover.

When describing leaves, we used technical terms because they are much more precise in the description of leaf shape and arrangement. The plant anatomy illustrations in the back of the book show leaf shapes, leaf divisions, leaf margins, and leaf arrangements. The glossary also describes many of the technical terms. Within the cactus family, the leaf category is divided into pads and spines.

In flower descriptions, we described mostly flower arrangement and color and not necessarily the size and details of its reproductive parts. When length is given, it refers to a single flower unless another feature is specifically noted. For instance, in Grasses and Grasslike Plants, the length measurement refers to the inflorescence because the flowers are so tiny.

When describing fruit, we included only the fruit type unless more information is helpful to distinguish it from another plant.

Flowering season includes both flowering and fruiting time, but does not usually extend to include the second bloom that is typical in Arizona, which receives summer monsoon rains. The flowering season is specific to the Grand Canyon, although throughout each species' range, flowering time varies based on the weather patterns where it grows. We used four primary sources for the flowering season: 1) *Annotated Checklist of Vascular Plants of Grand Canyon National Park* (Phillips, Phillips and Bernzott, 1987); 2) Museum Collection Herbarium at Grand Canyon National Park; 3) Desert Botanical Garden Herbarium in Phoenix; and 4) Deaver Herbarium at Northern Arizona University.

The elevational ranges are specific to the Grand Canyon. The four sources listed above provided this information. We listed elevations in English units (feet) because most maps are referenced this way.

When we described similar species within a species account, we used **boldface** to indicate a species that is not described elsewhere in the text.

When we mentioned specific uses for a species, we focus on native people within the Grand Canyon region. We limited other details of use to the desert regions of the Southwest extending into California, Utah, Nevada, Arizona, New Mexico, Texas, and Mexico. By including medicinal uses for certain plants, we intend to provide the readers information that enhances their appreciation for the botanical world. We do not recommend the use of any plant internally or externally without the advice and guidance of a professional herbalist. Furthermore, a permit is required for collecting plants within Grand Canyon National Park.

Thumbnail Identification Guide

Small thumbnail photographs in this section are grouped by flower color and then by other flower features to help you identify wildflowers and find their description in this book. Cactus flowers are not included because they are easily identified as members of the cactus family. Each thumbnail includes the page number that the flower occurs on. Please keep in mind that a variety of colors may occur in a single species, some individual flowers present several colors, and judging colors is subjective. You may need to look in the pink section to find the page for the purple flower that you've found, or you may need to look in the white section for a flower that has a large yellow center surrounded by small white petals.

In the dry deserts of the Southwest, many plants flower for only a short time. Look to their main description to identify them when they lack flowers.

We created most of the thumbnails from images in the book, which are credited where they appear at full size. A few thumbnails are created from images that we were unable to fit in the book, and we thank those photographers on the contributor list whose images were used for this purpose.

WHITE Few Petals					
 p. 164–65	 p. 180–81	 p. 130–31	 p. 182–83	 p. 144–45	

WHITE Few Petals					
 p. 60–61	 p. 172–73	 p. 118–19	 p. 168–69	 p. 180–81	

WHITE Many Petals					
 p. 218–19	 p. 212–13	 p. 224–25	 p. 204–5	 p. 212–13	

Many Petals	WHITE Clusters of Small Flowers				
p. 196–97	p. 190–91	p. 92–93	p. 90–91	p. 196–97	

WHITE Clusters of Small Flowers					
 p. 128–29	 p. 142–43	 p. 134–35	 p. 134–35	 p. 140–41	

HOW TO USE THIS BOOK

WHITE   Clusters of Small Flowers

p. 140–41   p. 94–95   p. 112–13   p. 150–53   p. 142–43

WHITE   Clusters of Small Flowers   **WHITE   Unusual**

p. 88–89   p. 118–19   p. 86–87   p. 188–89   p. 166–67

WHITE   Unusual   **YELLOW   Few Petals**

p. 80–81   p. 148–49   p. 150–51   p. 170–71   p. 138–39

YELLOW   Few Petals

p. 138–39   p. 116–17   p. 116–17   p. 208–9   p. 132–33

YELLOW   Few Petals   **Many Petals**

p. 162–63   p. 88–89   p. 148–49   p. 168–69   p. 182–83

4   HOW TO USE THIS BOOK

YELLOW   Many Petals

p. 218–19   p. 220–21   p. 192–93   p. 222–23   p. 210–11

Many Petals | YELLOW   Clusters of Small Flowers

p. 212–13   p. 202–3   p. 138–39   p. 84–85   p. 76–77

YELLOW   Clusters of Small Flowers

p. 114–15   p. 138–39   p. 192–93   p. 208–9   p. 206–7

YELLOW   Unusual

p. 142–43   p. 96–97   p. 194–95   p. 158–59   p. 176–77

RED

p. 154–55   p. 132–33   p. 100–109   p. 86–87   p. 174–75

HOW TO USE THIS BOOK

RED

p. 176–77	p. 178–79	p. 152–53

LIGHT PINK

p. 90–91	p. 90–91

LIGHT PINK

p. 74–75	p. 220–21	p. 186–87	p. 57	p. 130–31

DEEP PINK

p. 174–75	p. 156–57	p. 120–21	p. 176–77	p. 136–37

DEEP PINK

p. 122–23	p. 214–15	p. 170–71	p. 70–71	p. 122–23

LAVENDER

p. 178–79	p. 164–65	p. 162–63	p. 228	p. 146–47

LAVENDER

 p. 226–27
 p. 136–37
 p. 204–5
 p. 96–97
 p. 110–11

LAVENDER

 p. 92–93
 p. 178–79
 p. 198–99
 p. 128–29
 p. 150–51

PURPLE

 p. 112–13
 p. 168–69
 p. 120–21
 p. 226–27
 p. 120–21

PURPLE | BLUE

 p. 144–45
 p. 156–57
 p. 226–27
 p. 122–23
 p. 164–65

BLUE

 p. 136–37
 p. 160–61
 p. 160–61
 p. 158–59
 p. 98–99

HOW TO USE THIS BOOK   7

Geologic Time and Significant Biological Events

Time	Eon	Era	Period	Epoch	Rocks of the Grand Canyon Region	Significant Biological Events in Geologic Time	Other Notes
0	Phanerozoic Eon	Cenozoic Era	Quaternary	Holocene / Pleistocene	San Francisco volcanic field	Humans / Ice Age	Age of the Mammals
2			Tertiary	Pliocene / Miocene / Oligocene / Eocene / Paleocene		Carving of the Grand Canyon	
66		Mesozoic Era	Cretaceous		Mesa Verde Group / Mancos Shale	Primates / Flowering plants dominate / Birds	Age of the Dinosaurs and the Cycads
144			Jurassic		Dakota Sandstone / Morrison Formation / San Rafael Group / Navajo Sandstone	Conifers present worldwide / Mammals	
208			Triassic		Glen Canyon Group / Chinle Formation / Moenkopi Sandstone	Cycad and ginkgo families / Conifers and ferns diversify / Dinosaurs	
245		Paleozoic Era	Permian		Kaibab Limestone / Toroweap Formation / Coconino Sandstone / Hermit Shale	Pangea / Rise of conifers, horsetail	Age of the Fish, and ancient life
286			Carboniferous — Pennsylvanian		Supai Formation	Mosses / Conifers, widespread tree forests / Insects evolve wings / Crinoids	
320			Carboniferous — Mississippian		Redwall Limestone	Seed ferns expand	
360			Devonian		Temple Butte Formation	Amphibians / Ferns and fern allies / Decline of trilobites / First seed plants	
408			Silurian			First vascular plants (liverworts)	
438			Ordovician				
505			Cambrian		Muav Limestone / Bright Angel Shale / Tapeats Sandstone	Fungi, insects, and fish / Trilobites / Multicellular marine algae / Arthropods	
544	Proterozoic Eon	Precambrian (88 percent of Earth history)			Grand Canyon Supergroup	First animals (marine protozoa, worms, jellyfish)	Multi-celled life
1.0							
1.6					Vishnu Schist and Zoroaster Granite metamorphic suite		Single-celled life
2.5	Archean Eon					Stromatolites	
3.8						Bacteria / Blue-green algae	No rock record
4.6					no rock record	Earth forms	

Time scale: millions of years ago (0–544); billions of years ago (1.0–4.6)

The geologic timeline shows events in the plant and animal worlds during the time when rocks in the Grand Canyon were forming. —Graphic © 2006 by Bronze Black

GRAND CANYON ECOLOGY

Grand Canyon Plant Communities

More than 1,750 different species of plants grow in Grand Canyon National Park, representing nearly half of Arizona's flora, which is the fourth most diverse state flora in the country. In part, this diversity is attributed to a wide range of habitats that occur in the region. The Grand Canyon provides an incredible variety of places where plants can grow, ranging in elevation from 1,200 feet at Lake Mead to over 8,800 feet on the Kaibab Plateau, encompassing everything from spruce-fir forest to Mojave desertscrub, with springs, seeps, and creeks, and a large river running right down the center. Further contributing to the diversity is the Grand Canyon's regional position, where three of the four North American deserts converge.

To make sense of complex landscapes, botanists and biogeographers group plant communities in a variety of ways, including biomes, biotic provinces, floristic provinces, life zones, and habitat types. In 1890, C. Hart Merriam developed the concept of life zones by observing how plant and animal communities change from the bottom of the Grand Canyon to the top of the San Francisco Peaks near Flagstaff, Arizona. While ecologists continue to refine this early concept to include additional factors such as precipitation, soil types, and temperature, plants remain the fundamental component that define a specific life zone or community. As Stephen Trimble, a freelance writer and photographer, describes in *The Sagebrush Ocean: A Natural History of the Great Basin*, "in the end, all these approaches produce maps with different labels... but their boundaries are drawn in approximately the same places time after time: there are recognizable differences in communities."

More than nine hundred species of plants, representing over half of Grand Canyon National Park's flora, have been documented in the Inner Canyon, which is everything below the canyon rims. This book covers the lower-elevation areas beginning below the piñon-juniper woodland and extending to the river, including desertscrub and riparian life zones.

Riparian communities include marshes, seeps, springs, and the narrow ribbon of lush, green, fast-growing trees and shrubs that occurs along the Colorado River and perennial side creeks. Marshes develop in side canyons and along the Colorado River in return channels and backwaters where the water moves slowly enough for fine sediments to deposit. Seeps and springs are abundant in the canyon, varying greatly in their size and discharge. Their vegetation also varies depending on the reliability of available water, the size of the area, the slope, and the exposure. Side canyons provide moist, shaded enclaves that are capable of supporting a large diversity of plants. Many side canyons in the Grand Canyon are dry for most of the year and support few riparian species, while other side canyons, such as the Little Colorado River Gorge and Havasu Canyon, are more open and have large, perennial streams. Some, such as Buck Farm and Saddle Canyons, have lush, cool, shady glens, which contrast strongly with the open, sunny slopes. All side canyons are subject to occasional severe flash floods.

Desertscrub describes the mixture of plants that occur within the Mojave, Sonoran and Great Basin Deserts. In the Grand Canyon, the desertscrub zone is widespread, ranging from the slopes above the river to the Tonto Plateau and Esplanade Formation. Plants in desertscrub zones tend to be arid-adapted, small-statured, and tolerant of heavy winds and extreme sun exposure. A diversity of grass species grows among the shrubs and forbs in all of these desertscrub zones. Although not well-represented in the Grand Canyon, areas where grasses are dominant are called *desert grasslands*.

Grand Canyon Life Zones

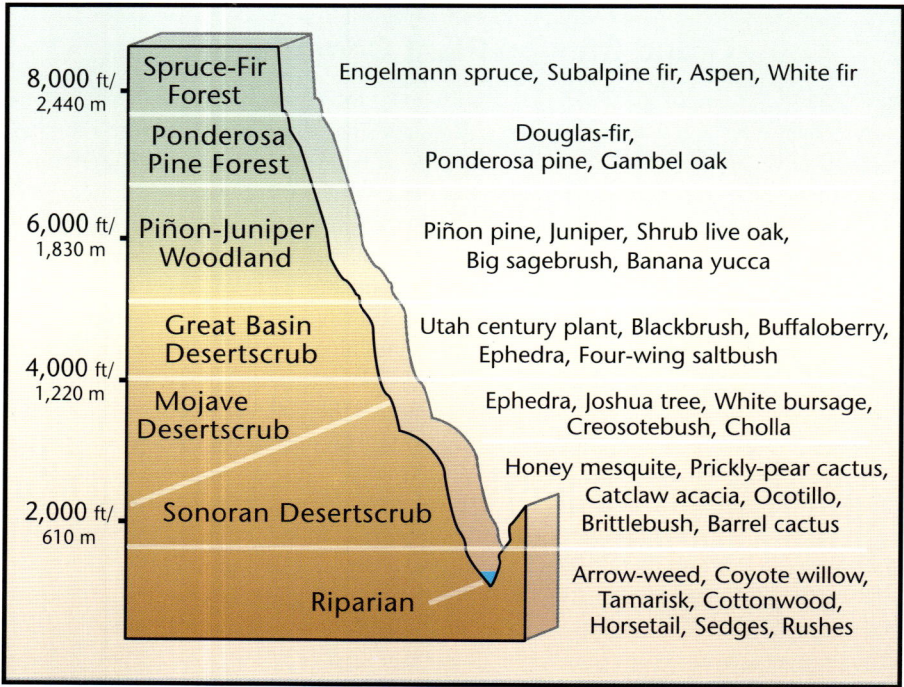

Each life zone lists dominant perennial species, yet plant distribution often overlaps into more than one zone. The slanted line that delineates Mojave and Sonoran desertscrub indicates that there is no distinct elevational boundary between these two zones in the Grand Canyon. —Graphic © 2006 by Bronze Black

While water and elevation are primarily responsible for the distribution of different kinds of plants in the Inner Canyon, the regional position of the Grand Canyon is also of critical importance. It is situated where three of the four North American deserts converge. To the north is the cold, high-elevation Great Basin Desert. To the west lies the Mojave Desert, a hot desert region. To the south and west is the Sonoran Desert, by far the most diverse of the four. Each of these three deserts reaches the limit of its range in the Grand Canyon. Although not extending into the Grand Canyon, the Chihuahuan Desert lies to the southeast. Some recognize the Painted Desert, which lies to the east bordering the Little Colorado River, as a fifth North American desert with a unique species assemblage.

In the Grand Canyon, we have the great privilege to observe a unique mingling of species from these deserts. Although the three deserts have distinct boundaries on maps, many plants are not exclusive to any one desert. The amount and seasonality of rainfall, daily temperature variation, and number of frost-free days help to define each desert. In Marble Canyon, plant communities display a mixture of cold-tolerant plants with Great Basin Desert influences. Species from both the Sonoran and Mojave Deserts prevail from the Little Colorado River to Grand Wash Cliffs, and there is debate as to which desert has the most influence because

North America is home to four major deserts, the Mojave, Sonoran, Great Basin, and Chihuahuan. —Graphic © 2006 by Bronze Black

many plants share allegiance to both deserts. Ocotillo (*Fouquieria splendens*) and California barrel cactus (*Ferocactus cylindraceus*) are definitely Sonoran, while white bursage (*Ambrosia dumosa*) and creosotebush (*Larrea tridentata* var. *tridentata*) inhabit both deserts.

There are at least three possible scenarios for plant dispersal in the Inner Canyon. Woodland plants can edge down slopes or wash down drainages. Great Basin and desert grassland plants can travel downstream from Glen Canyon and the upper Colorado River watershed. Mojave and Sonoran Desert species can disperse upstream along the river corridor.

The river corridor in the Grand Canyon is a well-known, low-elevation route for migratory birds, allowing them to avoid the climatic extremes of high elevations. It is reasonable to think that plants could similarly disperse along the narrow ribbon of warmer climate where water and resources are abundant, possibly with the aid of birds. The concept of upstream migration is not empirically obvious because the river flows downstream, counter to the primary direction of migration. For the river traveler, it may seem to be the first occurrence of a species, but in reality, these individuals represent the last outposts of plants moving upstream from warmer climates to the west. Intolerance to subfreezing temperatures is certainly not the only factor influencing Inner Canyon plants, but it is probably the most important.

Many familiar canyon plants experience this up-canyon migration. Beavertail cactus (*Opuntia basilaris*), a Sonoran Desert species, reaches to Lees Ferry and beyond. Brittlebush (*Encelia farinosa*), a desert shrub, reaches as far as River Mile 40 in Marble Canyon, and is a dominant member of the flora downstream to Grand Wash Cliffs. Ocotillo reaches its upstream limit on benches upstream of Havasu Creek and creosotebush reaches upstream to just below Fern Glen Canyon.

Although the changes to water flow by Glen Canyon Dam and the recent arrival of exotic plants have altered plant communities, the vegetation of the Grand Canyon has remained relatively stable for the last 8,000 years. Preserved fragments of twigs, seeds and leaves from ancient pack rat middens have allowed paleobotanists to reconstruct local floras. These middens show that a considerably different vegetation composition existed 20,000 years ago. Woodland species such as juniper (*Juniperus* species) and single-leaf ash (*Fraxinus anomala*) grew as much as 3,000 feet below their present lower-elevation limits, where they intermingled with desert species. The distinction between hot desert and woodland zones was less apparent in the past. As the past has shown us, there will once again be significant landscape-level shifts in species composition and diversity in the Grand Canyon.

—ARTHUR M. PHILLIPS III

The shrubs that form the pre-dam high-water line are immediately below the cliffs and above the riparian vegetation near the river. —Photograph by Lori Makarick/National Park Service

Riparian Vegetation Then and Now

The banks of the Colorado River in the Grand Canyon have not always displayed such a surreal contrast of colors. A dense green band of vegetation runs parallel to the tan slopes of desertscrub just above it. This thin strip of green plant life is called the *riparian zone*, from the Latin word *riparius*, meaning "of a riverbank." The riparian zone provides important food and shelter for insects, mammals, amphibians, reptiles, and birds, in addition to offering shade to canyon visitors. This riparian zone as we know it today is a product of the creation and operation of Glen Canyon Dam.

Our knowledge about pre-dam vegetation comes from a myriad of sources. Early photographs from explorers in the late 1800s show sparse plant life along the river. In the 1930s, botanists collected a limited number of specimens and described the species assemblage. Throughout both centuries, a host of canyon visitors described the landscape as they saw it in journals and other written documents.

Historically, spring runoff floods scoured and removed most riparian plants each year. High flows also buried plants with sediment, changed soil chemistry, and broke off limbs and leaves thereby damaging mature plants. Annual flooding benefited the plants as well by providing additional nutrients and naturally pruning diseased or dying branches, which resulted in vigorous new growth. Native riparian communities throughout the world are adapted to the effects of seasonal floods.

In 1963, Glen Canyon Dam was completed, providing water storage and hydroelectric energy for many western states. It also dramatically altered the downstream ecosystem. Currently the tempered, regulated flows never reach their historic spring level of 100,000 cubic feet per second. Over time, a thick band of riparian vegetation has begun to flourish along the banks of the Colorado River. This dense, new plant community is composed primarily of tamarisk (*Tamarix ramosissima*), willow (*Salix* species), seep-willow (*Baccharis* species) and arrow-weed (*Pluchea sericea*). Comparing it to a trip in 1938, botanist Lois Jotter Cutter observed in 1994 that, "the most obvious difference is the much greener look to the area immediately bordering the river, partly due to the great increase and frequency of the tamarisk. There have of course been changes in the numbers and distribution of individual species." It is unlikely the landscape will ever again look as it did on Cutter's 1938 trip. However, scientists and managers have identified the need for periodic high flows that mimic historic spring floods and are working to incorporate them into dam management.

In places above the river, a community of trees and shrubs still mark the pre-dam high-water line. To imagine the enormous amount of water that once flowed through the canyon during floods, look for a line of vegetation made up of netleaf hackberry (*Celtis laevigata* var. *reticulata*), catclaw acacia (*Acacia greggii*), western redbud (*Cercis orbiculata* var. *orbiculata*), western honey mesquite (*Prosopis glandulosa* var. *torreyana*), roundleaf buffaloberry (*Shepherdia rotundifolia*), and Apache plume (*Fallugia paradoxa*). This high-water community is now solely reliant on precipitation for survival, without which the plants decline and eventually die. Young plants of these species now establish closer to the river where their roots can reach water. This shift is yet another part of the changing ecosystem.

—MARISA HOWE

Desert Plant Adaptations

Deserts are extreme. Hot days, cool nights, dry air, and wind make survival for any creature a challenge. How do plants persist in such a harsh, unforgiving setting? Desert plants have adopted brilliant strategies to deal with these conditions. Modifications in physical structure prevent excess water loss and enhance water uptake from deep within the soil. Behavioral modifications help plants to function at times when it is most efficient to do so, often during the coolest times of day and wettest months in the season. Many plants combine strategies to maximize their efforts during periods of precipitation and optimal temperatures.

Plants that have altered their physical structure in order to better adapt to dry environments are called *xerophytes*. Cacti are a good example. They have thick, succulent stems that store excess water, allowing them to go long periods without rain, and waxy skin that seals in moisture. They have modified their leaves into spines so that less water is lost during transpiration, the process by which water evaporates from plants. Shallow, horizontal root systems allow them to gather surface moisture from a large area and absorb it rapidly during sporadic, heavy rains. One heavy rainfall provides enough water for some cacti to endure through several years of drought. Long, dense spines reflect the sun's rays, protect the plant from animals that would like to eat it, and provide shade from intense sunlight. These physical features allow cacti to flourish in hot, dry climates.

A more specialized group of xerophytes are the *phreatophytes*, or water-seeking plants. Their roots extend to incredible lengths in order to reach deep within the ground to find water. Creosotebush (*Larrea tridentata* var. *tridentata*) and western honey mesquite (*Prosopis glandulosa* var. *torreyana*) are two examples. With their extremely deep root systems, they seek out groundwater from deep beneath the surface that few other plants can access. This gives them a competitive advantage, allowing them to persist during periods of drought, which may kill other plants. The longest recorded mesquite roots reach as deep as 80 feet below ground.

Do plants have behavior? In the plant world, it is a bit different than what we typically think of as behavior, but how and when a plant's leaves, flowers, and fruit emerge and how it stores its reserves are considered plant behavior. Because of the extreme heat and dryness of the desert, plants function best during the coolest times of day and during the wettest months of the season. *Annuals* and *perennials* are two broad categories of plants that are grouped based on behavior—their strategy for survival and reproduction.

Annual plants grow from seed each year and bloom, set seed, and then die within a short period of time. Rather than expending precious energy to develop a root system that allows them to sprout every year, annuals invest all of their energy into seed production. They respond quickly to rainfall and can germinate, grow, mature, produce flowers and seeds, and die within a matter of weeks. Annuals are most abundant following periods of moisture. In the Grand Canyon, winter moisture allows seeds to germinate in the late winter and early spring months (February through May). Some annuals may sprout again during the summer monsoon rains (July through September). However, the seeds of annuals will not put forth the energy to germinate if the rains fail to come. They will remain dormant in the soil, waiting for sufficient precipitation to come the next year, increasing their chances for successful reproduction.

Perennial plants behave quite differently than annsuals. Instead of rushing to produce seeds in the tight window of a few weeks every year, they invest their energy into extensive root systems. This allows them to store energy beneath the ground and provide a foundation for sprouting each year. In the desert, perennials have the ability to become dormant if conditions are too harsh and may appear dead like ocotillo (*Fouquieria splendens*). Most of the year, this plant's thorny, gray mass of stems is hardly noticeable while it conserves its resources for growth and reproduction. Following a good rain, its leaves emerge, allowing it to capture more sunlight for photosynthesis, which fuels growth. Read more about each plant's unique strategy in the individual plant descriptions in this book.

—KRISTIN HUISINGA

Leaves of Ocotillo *Fouquieria splendens*
—Photograph © 2006 by Celia Southwick

Ocotillo (*Fouquieria splendens*) **with leaves** (*left*) **and without** (*right*)
—Photograph © 2006 by Geoff Gourley

Growing History: People and Plants in the Grand Canyon

As we look around the Grand Canyon, we see towering multihued cliffs, the mighty Colorado River, and an environment where plants cling to dizzying rock slopes, searching out spots with more moisture or less sun. Everywhere, we see life supremely adapted to its place and role in the canyon ecosystem, a place we define as the very essence of a wilderness. What may not be obvious at first glance, however, is the presence of people in this ecosystem.

Our view of the environment around us is strongly influenced by our underlying worldview, and for most in the United States, this view has been shaped by the history of colonial expansion into what was seen as a pristine, untamed "wilderness." Contrast this view with that of the indigenous people who lived and evolved within this wilderness. Not surprisingly, the views often fundamentally diverge. Indigenous cultures rarely include a concept of wilderness because people are part of the natural world, not separate from it. Any view about the land and what is natural is a cultural construct.

For the past 10,000 years, people have called the Grand Canyon home. They planted and cultivated food crops, including domesticated plants. Many domesticates, such as corn, beans, squashes, gourds and cotton, vanished from the landscape when people left and no longer tended them. Even though we rarely see these well-known crop plants growing wild in the Grand Canyon, their presence during earlier times is clearly recorded in archaeological deposits. A trip to Supai, where the Havasupai Indians have lived for at least seven hundred years, will quickly link the past to the present, demonstrating that domesticates are quite alive and well in the Grand Canyon.

While many well-known domesticated plants have their origins in other places, some native Grand Canyon species may have also begun the journey from wild to domesticated. The point at which a species becomes domesticated is not necessarily clear-cut. The change is accomplished in incremental steps, each succeeding generation of plants incorporating additional human-selected traits. For food plants, this might be a sweeter taste, larger fruits or roots, or fewer spines. For others, it might be longer fibers, more flexible shoots, better healing capabilities, or the ability to survive during droughts. People noted subtle properties in a given species and selected those individuals or populations that best fit a certain need. When found, such plants would be intentionally given a competitive edge by watering, selective weeding, or propagating. In essence, this was genetic engineering in its most basic form and the start of the domestication process.

Multitudes of native species are used as food, fiber, medicine, dye, craft materials, and for a myriad of other daily services. Among the most useful are the agaves (*Agave* species). The heart, or base, of the plant provides a sweet food that can be harvested, easily transported, and readily preserved. The leaves provide fiber, the stalks can be fashioned into tools, and both have medicinal value. They are ideally suited to the domestication process because they can reproduce vegetatively. This cloning ability means that desirable traits in a single plant can be reproduced easily, without the need for planting seeds and then waiting to see if the new plants have the desired qualities. The common Grand Canyon species, Utah century plant (*Agave utahensis*), was propagated this way, but to what degree is not known. Perhaps the clearest evidence of human-plant interaction survives in the form of a unique agave species, Grand Canyon century plant (*Agave phillipsiana*), first formally named and described by botanist Wendy Hodgson. To date, this species has been found only near archaeological sites and is unrelated to any other local agave species. It appears to be a direct relic of human activities, and populations of this plant may be declining in their absence.

Two other fiber-producing plants, yucca (*Yucca* species) and dogbane (*Apocynum cannabinum*), also possess potential for human modification because they reproduce vegetatively and are extremely useful. Longer stems and leaves yield longer fibers, which aid in efficient textile production.

Potentially, the greatest lasting effect of the indigenous people is not the genetic manipulation of individual species but the management of entire plant communities through habitat modification, selective harvesting, and seed scattering. People cleared areas by burning, dug and planted fields, harvested materials, removed undesirable plants, and trimmed branches or cut plants to ground level. They spared the western honey mesquite trees (*Prosopis glandulosa* var. *torreyana*) with the best-tasting pods from use as firewood. When clearing fields, they left and even encouraged edible native greens such as prince's-plume (*Stanleya pinnata* var. *pinnata*) or seed species such as Indian ricegrass (*Achnatherum hymenoides*). When native people encounter desert tobacco (*Nicotiana obtusifolia* var. *obtusifolia*) in the Grand Canyon, they collect some ripe seed and broadcast it around the area, an almost subconscious action to see this important plant flourish. Through time, these subtle pressures modify the landscape, both at a community and genetic level.

The plant communities we see today in the Grand Canyon are the result of generations of coevolution between people and their environment. It has only been within our recent history that these lands have been viewed as wilderness—a place where the forces of nature prevail and the human imprint is relatively unnoticeable. When hiking up a side canyon, perhaps we can picture ourselves 3,000 years earlier. Rather than seeing the shadow cast by an agave stalk on sun-dried Indian ricegrass as a photographic opportunity, we might see our farming activities culminating in a meal for our family. With understanding and appreciation, we can preserve the reciprocal relationship that indigenous people had with the landscape.

—MICHAEL YEATTS

Grand Canyon Century Plant *Agave phillipsiana*
—Watercolor © 2006 by Lora Colten

Banana Yucca *Yucca baccata*
—Photograph © 2006 by Bob Rink

The Foundation of Life:
Algae and Macrophytes of the Colorado River

In the Colorado River, algae and macrophytes are the base of the food web for aquatic organisms, which in turn are eaten by other insects, birds, and insectivorous vertebrates. Algae are plantlike organisms that lack root systems and range from microscopic size to several meters long. Macrophytes are aquatic plants large enough to be seen by the naked eye that grow both below and above the water surface, either with or without roots. Algae and macrophytes convert the sun's energy into available food in the form of carbon for other lifeforms.

Macrophytes and algae also provide critical habitat; their branches and filaments offer protection for aquatic invertebrates and fish. Waterweed (*Elodea* species), a macrophyte, has roots to anchor it into riverbed sediment, and cladophora (*Cladophora glomerata*), an alga, uses holdfasts to attach to submerged cobbles and boulders. In general, algae and macrophytes are less of a food source than the diatoms, or single-celled algae, that grow on their surface by the millions and produce high calorie lipids. Microscopic invertebrates, another vital food source for young fish, also depend on the algal communities for growth and reproduction.

Prior to the construction of Glen Canyon Dam, the food base of the muddy Colorado River was primarily found on plant debris that washed in from its numerous tributaries. Under such conditions, diatoms grew on floating branches and logs, where invertebrates easily consumed them. Today, Glen Canyon Dam traps the rich, food-laden debris and sediment. Downstream of the dam, the much-clearer, nutrient-deprived water permits photosynthesis beneath the surface for much of the year. In this setting, algae and macrophytes have become more abundant and are now the primary food base. Many native invertebrates did not adapt to the change in the food base and water temperature, and alien species invaded the cold, clear water. Glen Canyon Dam creates a constant ecological disturbance from which the native aquatic life cannot easily recover.

—EMMA BENENATI and JOSEPH SHANNON

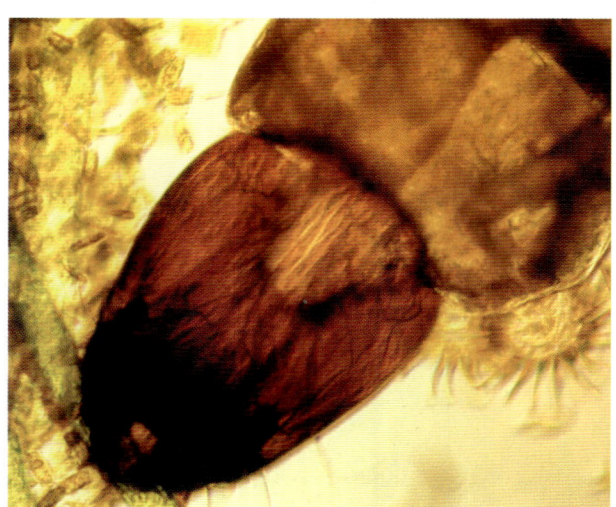

Midge eating diatoms from a filament of algae
—Photograph © 2006 by Dean Blinn

Biological Soil Crust

Many visitors to the arid deserts hear warnings about walking on soil crust and wonder, "what is this stuff and why is it so important?" At first glance, soil crust appears to be a lifeless coating on the soil's surface, but it is actually a diverse community of living organisms such as algae, cyanobacteria (blue-green algae), bacteria, lichens, mosses, liverworts, and fungi. You may also hear it called cryptogamic, microbiotic, cryptobiotic, and microphytic, which refer to different types of biological soil crusts.

The organisms within biological soil crusts decompose and mix with soil particles and other organic materials to form a rich and valuable resource for ecosystems. As a community, they fulfill an important ecological role. Crusts help to maintain soil stability, prevent erosion, contribute nutrients for plant growth, fix atmospheric nitrogen into a usable form, retain water in the soil for plants, and provide a comfortable bed for seedling germination.

Biological soil crusts are prominent in semiarid and arid environments throughout the world. They often cover areas not occupied by other plants and can represent as much as 70 percent of the living cover. Sparse plant cover in arid ecosystems leaves the soil vulnerable to erosion, and limited nutrients are slow to accumulate. Where crusts are well intact, they absorb some of the rain, which characteristically falls over a short period of time in southwestern arid lands. They also help decrease soil lost to wind. Biological soil crusts dominate slopes and terraces throughout the Grand Canyon. You may notice that areas lacking them tend to have more erosional gullies.

Recreation and grazing impacts fragile biological soil crusts, and once disturbed, they may take ten to one hundred years to recover. Near Lava Falls, where Robert Brewster Stanton's tripod stood more than one hundred years ago, the landscape still bears three imprints in the biological soil crust, illustrating its slow ability to recover. When crusts are damaged, plant productivity and diversity suffer, further affecting the larger ecosystem. Many seeds of native plants require biological soil crust for germination and growth. Without it, non-native plants are likely to colonize the disturbed, compacted ground. Hum this ditty to recall the importance of biological soil crusts: the planet Earth cries when biological soil crust dies.

—KATE WATTERS

Cryptobiotic soil crust
—Photograph by Lori Makarick/National Park Service

Close-up of mosses, lichens, and other organisms found in soil crusts —Photograph © 2006 by David Edwards

Hot Spots of Diversity: Springs in the Grand Canyon

Springs and seeps are among the rarest, yet most biologically diverse and productive ecosystems in the Grand Canyon region. They are remarkable hot spots of biodiversity particularly when their species diversity is contrasted against that of the surrounding landscape. While less than 1 percent of the Grand Canyon's area is composed of spring ecosystems, they support approximately 10 percent of the 1,750 plant species found here. Aquatic and terrestrial invertebrates, such as true bugs, beetles, and minute flies, abound at springs. In our arid region, amphibians, birds and mammals rely on spring ecosystems for food, water, and shelter. Grand Canyon springs also support endemic and rare species including Grand Canyon flaveria (*Flaveria macdougallii*), the tiger beetle (*Cicindela hemorrhagica arionae*) and the masked clubskimmer dragonfly (*Brechmorhoga pertinax*).

Springs in the Grand Canyon fall into two ecological categories, those that emerge from hillslopes and cliff faces and those that originate in tributary channels. On hillslopes, springs are not subjected to scouring floods and tend to have stable vegetative cover with more plant and landsnail species. Springs that emerge from tributary channels are often scoured by floods. They support fewer species and have a dynamic, changing plant cover that is often dominated by weedy species adapted to the disturbed habitat.

In addition to the high level of biodiversity they support, springs are important relict ecosystems, places where some species have retreated as the climate has changed over geologic time. Springs provide this important ecological and evolutionary refuge to plants that are rare or at the extreme edge of their range. Outside of the Grand Canyon and other protected areas in northern Arizona, more than 90 percent of springs have been obliterated or significantly impacted by human activities. In the arid Southwest, water supplied by springs is a commodity. Consequently, most of the West's springs have been severely altered or eliminated by diversion, manipulation, and groundwater withdrawal. Even within Grand Canyon National Park, a pipeline transports water from Roaring Springs in Bright Angel Canyon to supply water for use on the North and South Rims. Exotic species also threaten the ecological integrity of springs as plants like tamarisk (*Tamarix ramosissima*) continue to invade. Recognizing the biological importance of springs is the first step to long-term protection and restoration of these ecosystems.

—LAWRENCE E. STEVENS

Cardinal Monkeyflower (*Mimulus cardinalis*) **and Watercress** (*Rorippa nasturtium-aquaticum*) **at Deer Creek Falls, which is fed by several springs** —Photograph © 2006 by Geoff Gourley

Seeking the Sweet Reward: A Pollinator Vignette

A springtime hike into Havasu Canyon is alive with the buzz and activity of hummingbirds, hawk moths, bees, and flies visiting a wide array of wildflowers. Their heads, breasts, and wings are often dusted with pollen grains as these pollinators seek the sweet rewards of nectar and pollen. Some of them will transport this living treasure to other flowers, serving as sexual liaisons that often go unnoticed as you travel into the canyon. Without pollinators, few plants would bear fruit, a staple in the diets of many wildlife species.

At the canyon rim, with their bright orangish red flash, penstemon (*Penstemon* species) and Indian paintbrush (*Castilleja* species) survive despite growing between slabs of slickrock. The broad-tailed hummingbirds (*Selaphorus platycercus*) that zoom by at high speeds can spot these flowers from afar, even remembering a sequence of patches that leads them to nectar rewards. On their spring migration up the Colorado River, these hummingbirds may have followed the seasonally progressing bloom of ocotillos, whose tubular flowers contain abundant nectar. Hummingbirds are incredibly allegiant to the same patches of flowers from year to year, sometimes visiting the same shrub several flowering seasons in a row.

Descending deeper into the canyon, you can feel the sharp leaves of yuccas (*Yucca* species) and agaves (*Agave* species) poking at your shins, ankles, and toes. When their reproductive urge overpowers the tendency to conserve water and energy, they shoot up tall stalks loaded with lovely, delicate blossoms. This might occur every few years, or during extended drought, it may be postponed for up to twelve years before a big flowering season occurs. While yuccas are only pollinated by yucca moths (*Tegeticula* species), agaves are visited by nearly any flying animal that is in need of nectar: bees, hummingbirds, and other nectarivorous birds and bats. Yuccas practice a mutualistic relationship, allowing their symbiotic moth larvae to feed on a percentage of the developing seeds in exchange for high levels of pollination. Because agaves are less selective of their floral visitors, the number of seeds they produce varies greatly from year to year.

Moth visiting flowers
of Desert Pincushion
Chaenactis stevioides
—Photograph © 2006
by David Inouye

At the canyon bottom, sacred datura (*Datura wrightii*) makes a dramatic appearance, unfurling its trumpet-like blossoms at sunset when hawk moths become active. The hawk moths pick up the scent of the datura blossoms wafting in the breeze, then hover in front of them, extending their long proboscises down into the floral tube to suck up nectar. They often continue feeding until dawn.

The canyon bottom is filled with plants such as cultivated gourds and devil's claw (*Proboscidea parviflora*), which attract daytime pollinators like solitary bees and honeybees. Alien to the Americas, honeybees naturalized in the Grand Canyon well over a century ago and now compete with native bees for floral rewards. Native bees have adapted by visiting flowers much earlier in the day than honeybees, thereby accomplishing the lion's share of the pollination work.

As the sun descends behind the cliff faces, you may hear one last *zoom!* It is a hummingbird whose identity is too hard to detect in the diminishing light, but one that loves the colorful blossoms of garden species introduced by canyon dwellers. Even where humans have shaped the flora of the Grand Canyon, native pollinators eagerly seek gifts of nectar and pollen.

—GARY PAUL NABHAN

Bee dusted with pollen from flowers of California Barrel Cactus *Ferocactus cylindraceus*
—Photograph © 2006 by Kathy Darrow

Hawk moth gathering nectar with a long proboscis
—Photograph © 2006 by David Inouye

Biodiversity in Peril: Exotic Species in the Grand Canyon

The Grand Canyon is loved and cherished both by those who walk along its trails and those who have only seen photographs. It has been set aside for future generations of humans and nonhumans alike and deserves the highest degree of protection. National parks preserve some of the best examples of native, functioning, healthy ecosystems.

A well-known hiking destination for river travelers, Saddle Canyon is an excellent example of a biologically diverse side canyon. One of the most striking aspects of this canyon is the beauty and diversity of the flora, which ranges from desertscrub to lush riparian vegetation. Everything appears calm while hiking on the trail, but an invasion is underfoot. Native plants rise like buoys above a sea of exotic brome grasses. Ripgut brome (*Bromus diandrus*) and red brome (*B. rubens*) have spread so rapidly in the past two decades that they now dominate whole slopes of the Inner Canyon, competing with native plants for water, sunlight and other resources. In such abundance, these grasses readily spread fire across vast acreage in desertscrub ecosystems. These zones are not adapted to fire, which often kills long-lived, native species such as prickly-pear cactus (*Opuntia* species) and western honey mesquite (*Prosopis glandulosa* var. *torreyana*).

Imagine the landscape of Saddle Canyon covered in nothing but exotic grasses. The uniqueness and diversity of this beautiful side canyon would be lost. Now picture this happening throughout the Grand Canyon or even across the Southwest. The exotic species and the mechanisms of invasion may not always be the same, but the end result is sadly familiar. Ecologists warn that exotic species' invasions will soon surpass habitat loss and fragmentation as the most devastating threat to biological diversity worldwide.

Exotic species, also referred to as *non-native* or *alien*, are those that have been moved, either intentionally or unintentionally, outside of the area where they naturally evolved. Throughout history, humans have transported and cultivated some plants to supplement naturally occurring food plants. They also transplant some species to new areas just for their beauty. Increased travel, both within the United States and among continents, helps accelerate the mass movement of plants. Some species hitchhike on wind, water, shoelaces, fur, or vehicles.

Those exotic species that are able to adapt, persist, and spread rapidly after arrival are called *invasive*. Of the more than 1,750 plant species in the Grand Canyon, approximately 10 percent are exotic, but not all of these have the potential to cause the widespread ecological change that invasives can. While it may be too late to address some of the more widespread invasive species like red brome, we would be remiss not to take notice of new species that are just now beginning to creep their way into the Inner Canyon. Russian olive (*Elaeagnus angustifolia*) is one such example. This exotic tree was originally imported for its beauty and ability to stabilize soil along streambanks and ditches. An aggressive colonizer, it is now dominant in southwestern riparian areas but is conspicuously lacking in the Inner Canyon. Biologists and volunteers have vigilantly removed young plants from there before they have established and spread.

Sometimes we can prevent invasive species from gaining a foothold, but other times it takes extensive human effort to restore native ecosystems and keep the onslaught of exotic species at bay. Riparian areas are one of the highest priorities for restoration and protection because they represent less than 1 percent of the landscape in the Southwest and are critical for supporting biological diversity. More than half of the southwestern bird species breed in riparian habitats and nearly 65 percent of vertebrate species use them during different phases of life.

In an effort to return the riparian landscape at Lees Ferry to its native condition, biologists initiated a project to restore 10 acres dominated by tamarisk (*Tamarix ramosissima*). In 2001, the entire site was bulldozed to remove these aggressive exotic trees and herbicide was applied to sprouts. Approximately 1,000 native plants, including Fremont cottonwood (*Populus fremontii*), four-wing saltbush (*Atriplex canescens*) and willows (*Salix* species), were planted from locally collected cuttings and seeds and irrigated for the first three growing seasons. Monitoring of the site shows that 95 percent of the native plant species survived the first four growing seasons, with some trees even reaching 22 feet tall. Other native plants also colonized naturally from nearby populations. While initially displaced during the bulldozing, birds and mammal species have reinhabited the newly restored riparian forest, which has also attracted species not previously found in the dense thicket of tamarisk.

These extreme efforts may seem futile to some; many people argue that nature will take its course. In the long term, some species will be lost to extinction and other species will replace them. However, in the short term, if inaction persists, invasion by exotic species will accelerate the loss of diversity. We have the opportunity to protect and maintain our natural environments and the biodiversity that is important for survival of all life. Being active participants in slowing the onslaught of exotic species' invasions is a good place to start.

—DAN HALL, LORI MAKARICK, and FRED PHILLIPS

Top: **Lees Ferry restoration site before clearing tamarisk** —Photograph © 2006 by Fred Phillips Consulting

Bottom Left: **Lees Ferry restoration site cleared and newly planted** —Photograph © 2006 by Fred Phillips Consulting

Bottom Right: **Lees Ferry restoration site three years after planting** —Photograph © 2006 by Kristin Huisinga

Treasures of Diversity: Rare Plants in the Grand Canyon

What does it mean to be a rare plant? Is a plant rare if it is abundant within the walls of the Grand Canyon, but rare in other parts of Arizona or the Southwest? Ecologists and botanists group rare plants into three categories: 1) plants considered widespread geographically and locally rare, 2) plants restricted in overall range but locally abundant, and 3) those geographically restricted and locally sparse. The Grand Canyon harbors plants in each of these categories. Plants in categories 2 and 3 are also called *endemic plants*, meaning they are limited to a very specific geographical area, either abundant or not, such as those that are only known from the Grand Canyon.

Grand Canyon National Park hosts about 30 percent of the 290 Arizona plant species that various federal and state entities categorize as endangered, threatened, sensitive, rare, or endemic. Many regional botanists agree that the Grand Canyon might be a hotbed of speciation due to its unusual geology and relative isolation, two factors that often contribute to the evolution of unique plant characteristics. The high number of rare plants in the Grand Canyon is a direct result of the diversity of available habitats including riparian areas, subalpine forests, high deserts, low deserts, and wide sandy washes. A great deal of taxonomic work is still needed to investigate the peculiarities of many plants in the Grand Canyon, but the remote aspect of the vast landscape and limited funding and personnel make it difficult.

While the Grand Canyon may appear wild and untouched, many rare plants and their habitats have been lost or directly impacted by the proliferation of exotic species and the lack of seasonal flooding that followed Glen Canyon Dam. For instance, rushlike scurf-pea (*Psoralidium junceum*) was likely more common on sandbars along the Colorado River before the dam began holding back the large sediment-rich flood waters that once maintained these sandbars.

Grand Canyon also provides a refuge of protected habitat, especially along springs and streams. Satintail (*Imperata brevifolia*), a grass, and sawgrass (*Cladium californicum*), a sedge, are locally common in the Grand Canyon but rare in the Southwest because human activities have altered or destroyed many other riparian habitats.

You can help protect rare plants by treading lightly on the land, especially in such fragile environments as hanging gardens and perennial waterways. Leave the plants, flowers, and seeds in place so they can produce new generations and spread naturally. Resist the temptation to take just a few seeds home; this seemingly small act can impact rare plant populations. To safeguard rare beauties, we will not provide their specific locations in this book, but lucky travelers may be blessed with the opportunity to enjoy them.

—KATE WATTERS, LORI MAKARICK, and KRISTIN HUISINGA

FERNS AND FERN ALLIES

Originating over 400 million years ago during the Devonian Period, ferns and fern allies (horsetails, club mosses, spike-mosses, and others) are some of the earliest vascular land plants in the fossil record. Ferns differ from flowering plants in that they reproduce by spores housed in clusters, or *sori*, hidden on the undersides of the leaves, called *fronds*. Fronds and roots arise from horizontal, often underground, stems, called *rhizomes*, which are swollen with food reserves. The spore cases of mature plants dry out, expand, and contract, catapulting spores into the air. A spore lands on moist ground and develops into a *prothallus*, a small flat structure that contains male and female cells from which develops a fern. Fern allies employ a similar reproductive strategy, but instead of generating spores on the underside of their leaves, they bear spores in cone-shaped spikes atop stems or in leaflike organs in between their leaves.

While the diversity of flowering plants now exceeds that of ferns, their early reproductive strategy allowed them to endure through massive catastrophes and extinctions throughout geologic history. The success of their simple reproductive strategy is obvious; today, more than ten thousand species of ferns are found worldwide. As many as thirty-five species of ferns and other plant fossils are preserved in the Hermit shale in the Grand Canyon, a sedimentary layer that emerges at river level near Badger Canyon.

Grand Canyon is host to nearly twenty species of ferns, which belong to four different families. The ferns described here are all members of the maidenhair family, which is well represented in dry areas worldwide.

Horsetail, Scouring Rush *Equisetum x ferrissii*
HORSETAIL FAMILY (Equisetaceae)

Plants: Annual or perennial herb, with jointed stems, to 1 meter tall. **Leaves:** In crownlike whorls, reduced to scales. **Sori:** In terminal strobilus that are 10 to 25 millimeters long. **Spore-Producing Season:** January to July. **Elevation:** 2,100 to 5,200 feet.

Horsetails belong to an ancient group of plants whose ancestors were dominant during Jurassic time when dinosaurs roamed the landscape. Horsetail forms a band of dark green vegetation adjacent to the water, often mingling with arrow-weed (*Pluchea sericea*), tamarisk (*Tamarix ramosissima*), and coyote willow (*Salix exigua*). It has hollow, jointed, strawlike stems whose minute parallel ridges are rough to the touch. The photosynthetic stems bear tiny, black, deciduous scales at the joints, which are the evolutionary origins of leaves. Horsetail produces spores, a reproductive strategy shared with ferns.

Our dominant horsetail species produces only sterile spores and is a hybrid between two other Grand Canyon species, a perennial (**Equisetum hyemale**) and an annual (**E. laevigatum**). Being unable to reproduce sexually has not hindered it from establishing large colonies; it relies on shoot production and the ability to root from fallen stem joints. These adaptations enable it to thrive in constantly disturbed areas like riverbanks.

The lone genus in the family Equisetaceae, *Equisetum* boasts fifteen species worldwide in temperate and tropical climates. Our species occurs throughout the United States, reaching north to Canada and south to Mexico. In the Grand Canyon, it grows the length of the river and in side canyons with its parent species extending to the North Rim.

The roots provided silica-rich food and medicine to Native American people and early settlers, who also used the stems for scouring and polishing pots and pans, hence the common name *scouring rush*.

Maidenhair Fern *Adiantum capillus-veneris*

MAIDENHAIR FAMILY (Pteridaceae/Polypodiaceae)

Plants: Rhizomatous, perennial fern. **Fronds:** Long-petioled, bipinnately compound with cleft to lobed delicate segments, shiny green, 20 to 40 centimeters long. **Sori:** Along margins, hidden by folded-under tips of segment lobes. **Spore-Producing Season:** May to December. **Elevation:** 1,200 to 6,800 feet.

River travelers will inevitably notice the delicate fronds of maidenhair fern and their distinct sweet tea fragrance. This plant graces practically every seep, spring, and perennial waterway in the Grand Canyon. Often, the blossoms of cardinal monkeyflower (*Mimulus cardinalis*) or golden columbine (*Aquilegia chrysantha*) protrude from a thick tapestry of maidenhair fern.

The shiny green fronds of maidenhair fern are water-resistant, providing an adaptive advantage for inhabiting dripping cliffs, ledges, and seeps along streams, rivers, and canyon walls. Look for it tucked in crevices beneath even the most powerful waterfalls, like Deer Creek Falls. It is widespread, found in tropical and warm temperate regions of the Old and New Worlds. This plant prefers alkaline soils derived from limestone or other calcium-rich rocks. In the Grand Canyon, it is found from Marble Canyon to Grand Wash Cliffs and in scattered seeps on the South Rim.

The genus name, *Adiantum*, is Greek for "unwettable" in reference to its water-resistant leaves, and the specific epithet, *capillus-veneris*, means "Venus's hair," alluding to the pubic hair of the goddess of love. Havasupai people use the shiny, purplish black petiole in their exquisite coil baskets.

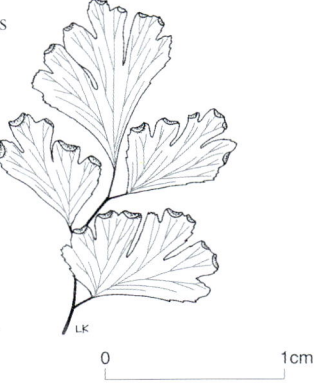

Maidenhair Fern
Adiantum capillus-veneris
—Illustration © 2006 by Lisa Kearsley

Horsetail *Equisetum x ferrissii*
—Photograph © 2006 by Arizona Raft Adventures

Maidenhair Fern *Adiantum capillus-veneris*
—Photograph © 2006 by Bob Rink

Scaly Cloak Fern *Astrolepis cochisensis* ssp. *cochisensis*
MAIDENHAIR FAMILY (Pteridaceae/Polypodiaceae) (*Notholaena cochisensis*)

Plants: Rhizomatous, perennial fern. **Fronds:** Long-petioled, once-pinnately compound, olive green, 11 to 30 centimeters long. **Sori:** Along veins, hidden by dense scales. **Spore-Producing Season:** June to October. **Elevation:** 2,000 to 7,000 feet.

Scaly cloak fern is relatively inconspicuous, but the long, narrow fronds with numerous, oval divisions bearing star-shaped hairs on their surface set it apart from the other three species described here. It prefers rocky slopes and limestone cliffs, occurring in California and Arizona, east to Texas and Oklahoma, and south into mainland Mexico and Baja California. In the Grand Canyon, it grows at Stanton's Cave (just downstream of South Canyon), Elves Chasm, and Boucher (at River Mile 97) and Havasu Canyons and is probably more widespread. The genus name, *Astrolepis*, is Greek for "star scale," referring to the hairlike scales on the underside of the frond segment.

Slender Lip Fern *Cheilanthes feei*
MAIDENHAIR FAMILY (Pteridaceae/Polypodiaceae)

Plants: Rhizomatous, perennial fern. **Fronds:** Long-petioled, bi- to tripinnately compound, pale green to green, 4 to 20 centimeters long. **Sori:** At tips, partially obscured by hairs. **Spore-Producing Season:** April to August. **Elevation:** 1,600 to 7,200 feet.

The rusty-colored, dense, scaly hairs on the lower surface are diagnostic of slender lip fern. The fronds top a wiry, dark brown stalk that is more than half the frond length. Slender lip fern is widespread, occurring from British Columbia, south to northern Mexico, and east to Virginia. In the Grand Canyon, it grows on cliffs, ledges, rocky slopes, and mesas usually in limestone or sandstone soils, extending from the river to the rims.

Spiny Cliff-Brake *Pellaea truncata*
MAIDENHAIR FAMILY (Pteridaceae/Polypodiaceae) (*Pellaea longimucronata*)

Plants: Rhizomatous, perennial fern. **Fronds:** Long-petioled, bipinnately compound, bluish green, 8 to 40 centimeters long. **Sori:** In marginal bands, hidden under frond margins that curve inward. **Spore-Producing Season:** April to August. **Elevation:** 3,250 to 5,000 feet.

The delicate, bluish gray hue of spiny cliff-brake contrasts with tawny cliffs where it is commonly found. This arid-land fern is a western species that grows in dry soils and rocky areas from Colorado and Texas west to California and south into Mexico. In the Grand Canyon, it grows in side canyons along the river, along Bright Angel Trail, and on the North and South Rims. The genus name, *Pellaea*, is Greek for "dusky," referring to the bluish gray leaves.

Scaly Cloak Fern *Astrolepis cochisensis* ssp. *cochisensis*
—Photograph © 2006 by Kate Watters

Spiny Cliff-Brake *Pellaea truncata*
—Illustration © 2006 by Bronze Black

Slender Lip Fern *Cheilanthes feei* —Photograph © 2006 by Kate Watters

GRASSES AND GRASSLIKE PLANTS

Southern Cattail *Typha domingensis*
CATTAIL FAMILY (Typhaceae)

Plants: Semiaquatic, rhizomatous perennial, to 4 meters tall. **Leaves:** Linear to 4 meters long. **Flowers:** In cylindrical, spikelike heads; male spikes above, 11 to 22 centimeters long; female spikes below, 7 to 35 centimeters long. **Fruit:** Microscopic achene. **Flowering Season:** April to October. **Elevation:** 1,200 to 5,500 feet.

Southern cattail, a globally widespread plant, most often grows in standing water. Its most notable feature is the chestnut brown, densely packed, cylindrical fruiting spike. Lacking these, it is still easy to identify. Southern cattail has long, thin, sword-shaped leaves that emerge from a stout stem, unlike the rounded or triangular stems of similar plants like sawgrass (*Cladium californicum*), sedge (*Carex* species), and rush (*Juncus* species). Prior to the construction of Glen Canyon Dam, cattail was uncommon along the river because the scouring of high-volume spring floods wiped out entire patches of mature plants. Today, they grow in dense populations in the slower-moving waters, creating marshes where they did not previously exist. Under current dam operations, river flows fluctuate each day, but because the thick rhizomes store water, cattails can persist for short periods without standing water.

Southern cattail occurs throughout the West, extending to the Gulf States of the Southeast and into Mexico. In the Grand Canyon, it thrives in fine, silty soils in scattered patches along the shoreline the length of the river and in side canyons with permanent water.

A less common species, **broad-leaved cattail (*Typha latifolia*)**, grows along the river only from Vasey's Paradise to the Little Colorado River and in scattered side canyons. Even when the male flowers have dispersed, a gap, up to 8 centimeters long, is visible between the male and female flowering spikes of southern cattail, which is lacking in broad-leaved cattail.

Whole, seemingly invisible, biological communities prosper within the depths of thick cattail stands. While these stands are proportionately small in relation to the total cover of riparian vegetation, they are vital to the Grand Canyon's biodiversity. They offer nest material, cover, and food for a wide array of invertebrates, amphibians, fish, mammals, and birds. You may see places where beaver clawed deep below the surface to acquire the tasty young shoots and stems.

Native people frequently used fire to manage overgrown marshes, thus increasing human accessibility to useful plants and wildlife. The long leaves of cattail were woven into mats, chairs, baskets, canopies, and clothing.

Gap between male and female spikes of Broad-Leaved Cattail
Typha domingensis
—Photograph © 2006 by John Running

Southern Cattail *Typha domingensis*
— Photograph © 2006 by Bob Rink

Indian Ricegrass

Achnatherum hymenoides

GRASS FAMILY (Poaceae/Gramineae)

(*Oryzopsis hymenoides*)

Plants: Perennial bunchgrass, 25 to 70 centimeters tall. **Leaves:** Narrow, involute, as tall as plant. **Flowers:** In openly branched panicles that are 5 to 20 centimeters long. **Fruit:** Grain. **Flowering Season:** April to August. **Elevation:** 1,200 to 7,900 feet.

Indian ricegrass grows in hearty clumps of stunning dark green in early spring and becomes straw-colored and brittle as the plant matures in the fall. Bearing spikelets at the end of wavy, capillary branches, the delicate stalks sway in the wind on hot summer days. Look for the dried flower stalks to recognize this plant long after its prime. Scan the riverbanks below Havasu Canyon for its taller Mediterranean cousin **smilograss (*Piptatherum miliaceum*)** with much smaller seeds and broader, flat leaves.

Indian ricegrass ranges from Mexico and Texas to central Canada across a broad range of habitats from deserts to piñon-juniper woodlands. In the Grand Canyon, plants inhabit sandy, dry, open areas the length of the river up to the rims.

Compared to other grasses, the large, plump seeds of Indian ricegrass are easy to harvest, very nutritious, and rich in protein—thus an important food source to wildlife and humans. Although not practiced by most today, Native Americans roasted the seeds, thus removing the dense white hairs while at the same time parching the seeds. Small mammals cache husked seeds, sometimes forgetting their locations, and the seeds germinate to form new populations.

Bushy Beardgrass

Andropogon glomeratus

GRASS FAMILY (Poaceae/Gramineae)

Plants: Perennial bunchgrass, to 1.5 meters tall. **Leaves:** Flat to folded, hairy-margined, 30 to 60 centimeters long. **Flowers:** In broomlike panicles that are up to 30 centimeters long. **Fruit:** Grain. **Flowering Season:** July to October. **Elevation:** 1,200 to 4,000 feet.

Along the banks of the Colorado River, the large plumes and wide leaves of bushy beardgrass turn rusty orange, announcing the arrival of fall. The many-leaved panicle clusters distinguish it from cane bluestem (*Bothriochloa barbinodis*), which has an essentially leafless flowering stalk. While usually taller than bushy beardgrass, **big bluestem (*Andropogon gerardii*)** is slender with a sparsely flowered, more open panicle.

Bushy beardgrass occurs throughout the United States, particularly in warmer climates. In the Grand Canyon, it occupies riverbanks throughout the length of the river and in side canyons with permanent water. You will likely notice it in the lower, warmer reaches of the canyon, where it is most abundant.

It is closely related to sorghum and sugarcane, both economically important crops. The genus name, *Andropogon*, is from the Greek words *andros*, meaning "man," and *pogon*, meaning "beard," referring to the long, white hairs on the male flowers of some species.

Indian Ricegrass *Achnatherum hymenoides*
—Photograph © 2006 by Kristin Huisinga

Smilograss *Piptatherum miliaceum*
—Illustration © 2006 by Kate Watters

Big Bluestem *Andropogon gerardii*
—Photograph © 2006 by David Edwards

Bushy Beardgrass *Andropogon glomeratus*
—Photograph © 2006 by David Edwards

Arizona Three-Awn *Aristida arizonica*
GRASS FAMILY (Poaceae/Gramineae)

Plants: Perennial bunchgrass, 15 to 100 centimeters tall. **Leaves:** Fine, strongly recurved, flat or with edges rolling inward, 3 to 30 centimeters long. **Flowers:** In sparsely flowered panicles that are 5 to 25 centimeters long. **Fruit:** Grain. **Flowering Season:** May to October. **Elevation:** 1,600 to 4,300 feet.

Three-awn is an easy grass to identify because of its distinctive three-parted awn that tops the flowers. The awns may assist seed dispersal in two ways: the windmill-like formation may help catch the wind or the three prongs may readily attach to animal fur. This grass flowers consistently each year, even in times of severe drought. Later in the year, the delicate bunches turn a soft, purplish hue as the leaves begin to dry and curl.

Arizona three-awn occurs in Colorado, Texas, Arizona, and northern Mexico on dry plains, open forests, and sandy or gravelly mesas. In the Grand Canyon, it grows on beaches, dry terraces, and rocky slopes. Arizona three-awn is one of the dominant grasses in the Inner Canyon, but despite its dominance, wildlife avoid it because of the sharp-pointed fruits.

Of the five species of three-awn found in the Grand Canyon, you will most likely see the two common perennial species, which have overlapping ranges. **Purple three-awn** (*Aristida purpurea*) has distinctly unequal glumes with up to 12-centimeter-long awns, as compared to the nearly equal glumes and shorter awns of Arizona three-awn. The small winter annual, **six-weeks three-awn** (*A. adscensionis*), has more densely packed panicles that begin flowering in February.

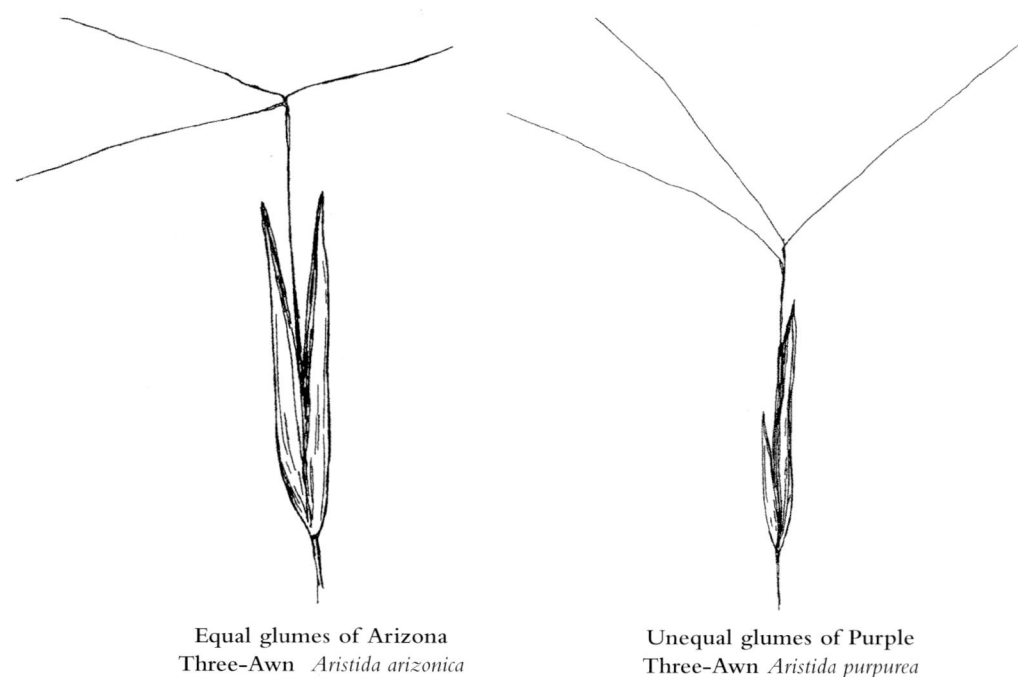

Equal glumes of Arizona Three-Awn *Aristida arizonica*
—Illustration © 2006 by Kate Watters

Unequal glumes of Purple Three-Awn *Aristida purpurea*
—Illustration © 2006 by Kate Watters

Six-Weeks Three-Awn
Aristida adscensionis
—Illustration © 2006 by Kate Watters

Arizona Three-Awn
Aristida arizonica
—Illustration © 2006 by Kate Watters

Cane Bluestem
GRASS FAMILY (Poaceae/Gramineae)

Bothriochloa barbinodis
(*Andropogon barbinodis*)

Plants: Perennial bunchgrass, to 1.5 meters tall. **Leaves:** Flat, 20 to 30 centimeters long. **Flowers:** In 5- to 14-centimeter-long terminal panicles, with clustered branches. **Fruit:** Grain. **Flowering Season:** April to October. **Elevation:** 1,300 to 3,400 feet.

Cane bluestem most commonly grows interspersed with other grasses, seldom forming dense stands. It can tolerate very dry soils, where other grasses cannot survive, and is also highly valued for its ability to control erosion.

A beardlike ring of dense hairs around each node distinguishes cane bluestem from look-alike species such as **silver beardgrass (***Bothriochloa saccharoides***)** and **little bluestem (***Schizachyrium scoparium***)**. Cane bluestem sometimes hybridizes with the closely related silver beardgrass, which has a smaller, narrower raceme and is less common.

Cane bluestem occurs from northern Mexico and California, east to Texas and Oklahoma, with geographically disjunct populations in South Carolina and Florida. In the Grand Canyon, it grows along the banks of the river from Lees Ferry to Three Springs, and in side canyons such as Buck Farm, Basalt, and Havasu Canyons.

The specific epithet, *barbinodis*, is from the words *barba*, meaning "beard," and *nodus*, meaning "joint," referring to the bearded nodes.

Sideoats Grama
GRASS FAMILY (Poaceae/Gramineae)

Bouteloua curtipendula

Plants: Perennial grass, 10 to 90 centimeters tall. **Leaves:** Flat, thin, hairy-margined, 2 to 25 centimeters long. **Flowers:** In one-sided racemes that are 5 to 30 centimeters long, with thirteen to eighty dangling spikelets that are up to 1 centimeter long. **Fruit:** Grain. **Flowering Season:** April to October. **Elevation:** 1,750 to 5,500 feet.

The flowering clusters of sideoats grama all dangle from one side of the tall stalk, a noticeable feature of all grasses in the genus *Bouteloua*. Sideoats grama grows in clumps and has curly leaves that turn purple and then straw-colored in the fall.

Sideoats grama is a highly adaptable and drought-resistant grass found throughout North and South America from Canada to Argentina. It is the state grass of Texas. In the Grand Canyon, sideoats grama grows on dry slopes and mesas from Lees Ferry to Grand Wash Cliffs. You may also encounter its dainty relative, the annual **six-weeks grama (***B. barbata***)** after monsoon rains, when its eyebrow-like, arching spikes carpet dry mesas and sandy slopes the length of the canyon. Another annual relative, **six-weeks needle grama (***B. aristidoides***)** also grows abundantly on slopes and mesas but is more common in the lower canyon. It lacks the eyebrow-shaped spikes.

Because of its extensive root system, sideoats grama can maintain its foliage during periods of drought, thereby ensuring a relatively consistent food supply for pronghorn, deer, and elk. Birds and small mammals relish the seeds.

Cane Bluestem
Bothriochloa barbinodis
—Illustration © 2006 by Lisa Kearsley

Silver Beardgrass
Bothriochloa saccharoides
—Illustration © 2006 by Lisa Kearsley

Little Bluestem
Schizachyrium scoparium
—Illustration © 2006 by Lisa Kearsley

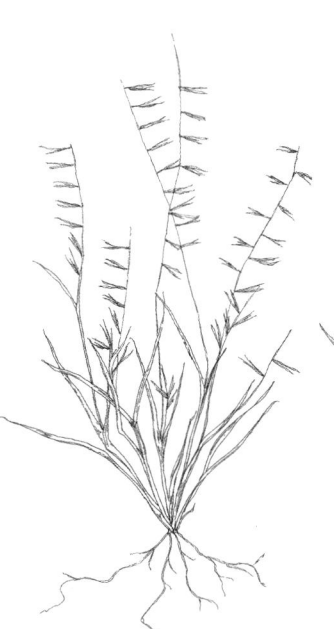
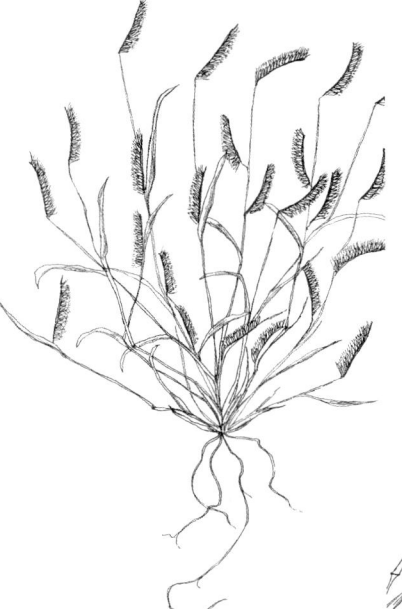
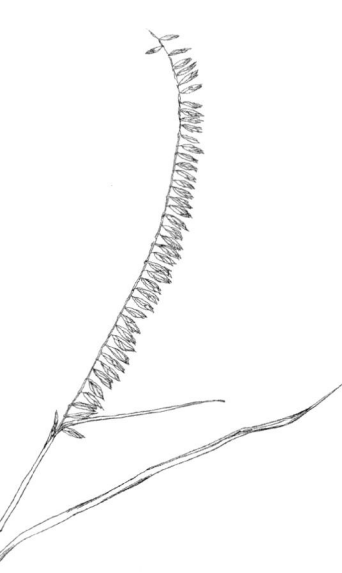

Six-Weeks Needle Grama
Bouteloua aristidoides
—Illustration © 2006 by Kate Watters

Six-Weeks Grama
Bouteloua barbata
—Illustration © 2006 by Kate Watters

Sideoats Grama
Bouteloua curtipendula
—Illustration © 2006 by Kate Watters

BROME GRASSES
Bromus species

GRASS FAMILY (Poaceae/Gramineae)

Cheatgrass *Bromus tectorum*

Plants: Exotic annual grass, 5 to 80 centimeters tall. **Leaves:** Flat or involute, hairy. **Flowers:** In drooping panicles that are 5 to 15 centimeters long. **Fruit:** Grain. **Flowering Season:** April to July. **Elevation:** 2,300 to 8,100 feet.

Red Brome *Bromus rubens*

Plants: Exotic annual grass, 10 to 60 centimeters tall. **Leaves:** Flat or involute, hairy. **Flowers:** In erect panicles, 3 to 10 centimeters long. **Fruit:** Grain. **Flowering Season:** February to August. **Elevation:** 1,200 to 6,800 feet.

Ripgut Brome *Bromus diandrus* (*B. rigidus* var. *gussonei*)

Plants: Exotic annual grass, 10 to 90 centimeters tall. **Leaves:** Flat or involute, long hairy. **Flowers:** In nodding panicles that are 7 to 15 centimeters long. **Fruit:** Grain. **Flowering Season:** April to June. **Elevation:** 1,700 to 6,800 feet.

Brome grasses are so widespread throughout the Grand Canyon that it would be hard to miss them. In general, the perennial species are native and the annual species are European or Mediterranean in origin. In the western United States, annual brome species cover enormous acreage and have degraded the ecosystems in which they thrive. Although fifteen brome grasses occur in the Grand Canyon, the three species described here are the most common and are all weedy annuals with hairy blades and sheaths, narrow spikelets, and rather long awns. The spikelets of cheatgrass and red brome have a purplish hue.

Cheatgrass occupies a wide variety of habitats in the northern regions of North America. It is very common on open slopes and is found in deserts, woodlands, and forests. It is usually found at higher elevations than the two other brome grasses.

Red brome is the most widespread of the three bromes found in the Inner Canyon. Throughout the year it blankets slopes and dunes, turning them green in early spring and reddish tan as it matures. The erect, tufted flowering head distinguishes red brome from the other two bromes.

Although overlapping in distribution, ripgut brome is much taller and stouter and is less common than red brome. It is easily differentiated by its dramatically long, rough awns that can injure mouths of foraging animals.

These species germinate in the fall and grow rapidly in the spring, often in dense patches. They require little precipitation to germinate. The roots penetrate up to 20 centimeters deep and branch out laterally, allowing them to thrive on dry sites with nutrient-poor soils. This combination of strategies affords them a competitive advantage over native species, particularly annuals. Brome grasses burn readily and carry fire through desert plant communities, which only rarely experienced fire prior to the invasion of exotic species. Frequent fire favors the persistence of brome grasses over native species.

One brome plant can produce fifty to three hundred seeds, which are viable for an average of two to five years. The seeds disperse by wind and attach to clothing and fur. Most of us are guilty of unknowingly transporting the seeds in our shoes and socks. Next time, carefully remove the seeds and discard them in the garbage rather than on the ground. The genus name derives from *bromos*, the ancient Greek name for "oat."

Cheatgrass *Bromus tectorum*
—Illustration © 2006 by Kate Watters

Red Brome *Bromus rubens*
—Photograph © 2006 by Lisa A. Hahn

Red Brome *Bromus rubens*
—Illustration © 2006 by Kate Watters

Ripgut Brome *Bromus diandrus*
—Illustration © 2006 by Kate Watters

Bermuda Grass *Cynodon dactylon*
GRASS FAMILY (Poaceae/Gramineae)

Plants: Exotic, creeping perennial grass, 10 to 80 centimeters long. **Leaves:** Flat to involute, 2 to 12 centimeters long. **Flowers:** In panicles, with two to seven spikelike, whorled branches. **Fruit:** Grain. **Flowering Season:** March to October. **Elevation:** 1,500 to 4,500 feet.

Bermuda grass forms prominent, bright green carpets on riverbanks in the Grand Canyon. In 1987, Bermuda grass was only known to occur downstream of the confluence of the Little Colorado River, and in just a few decades it has spread throughout the river corridor from upstream sources. This range expansion is partially due to the ability of its rhizomes and stolons to root in new locations once they are transported downstream. Although it is an exotic species, Bermuda grass anchors sandy banks along the river, thus retarding beach erosion.

Bermuda grass can be confused with scratchgrass (*Muhlenbergia asperifolia*) because of their similar spreading habits and leaf arrangement. A closer look at the flowering stalks shows the difference: Bermuda grass has fingerlike whorled branches, while scratchgrass stalks are fine and wispy. Even if no heads are present, long white hairs on the ligule identify Bermuda grass. Also with a hairy ligule and similarly rigid leaves, **inland saltgrass (*Distichlis spicata*)** must be in flower to set it apart. It has densely packed, many-flowered spikelets, not wagon spokes like Bermuda grass.

Native to Africa, Bermuda grass was introduced into the United States as a turf grass. Throughout its range, it grows on moist, saline soils. Bighorn sheep eat the young shoots and you may see them lazing about in dense patches of Bermuda grass along the river. The specific epithet, *dactylon*, is from the Greek *daktylos*, meaning "finger" or "toe," possibly referring to the fingerlike arrangement of the flowering branches.

Bermuda Grass *Cynodon dactylon*
—Illustration © 2006 by Kate Watters

Inland Saltgrass *Distichlis spicata*
—Illustration © 2006 by Kate Watters

Bermuda Grass (*Cynodon dactylon*) **covering the banks of the Colorado River**
—Photograph © 2006 by Geoff Gourley

Fluffgrass *Dasyochloa pulchellum*
GRASS FAMILY (Poaceae/Gramineae) (*Erioneuron pulchellum*)

Plants: Perennial grass, 2 to 15 centimeters tall. **Leaves:** Narrow, involute, 1 to 6 centimeters long. **Flowers:** In headlike panicles that are 1 to 4 centimeters long. **Fruit:** Grain. **Flowering Season:** March to October. **Elevation:** 1,200 to 5,000 feet.

Even as tiny seedlings fluffgrass is fun to encounter and easy to learn, although you might not recognize it as a grass. Dense fluff coats the small, linear leaf blades, helping to deflect sunlight and conserve water. Plants form small gray mounds crowded with short, dense spikelets. Fluffgrass may be small, but it is tough, remaining dormant for months, awaiting the right conditions to spring to life. In the late summer, the conspicuous moplike clusters carpet open areas in an otherwise stark, leafless landscape.

Fluffgrass, the only species in the genus *Dasyochloa*, is one of the more common grasses in the southwestern United States. It occurs from California east to Texas and has disjunct populations in Maryland. In the Grand Canyon, it occurs throughout the Inner Canyon, primarily on arid desert flats and rocky slopes.

Satintail *Imperata brevifolia*
GRASS FAMILY (Poaceae/Gramineae)

Plants: Perennial grass, to 1.5 meters tall. **Leaves:** Flat, 10 to 50 centimeters long. **Flowers:** In spikelike panicles that are up to 30 centimeters long. **Fruit:** Grain. **Flowering Season:** May to October. **Elevation:** 1,200 to 4,500 feet.

The silky-silvery flowering stalks of satintail, a water-loving desert grass, reach high above its leaves, which are nestled among other streamside plants. In the Southwest, development of riparian areas threatens this species because it requires moist sites—limited commodities in the desert. The Grand Canyon provides a refuge for satintail where it seems to be unusually abundant and thriving along the river corridor and some side canyons.

Satintail occurs from Texas west to California and south to Mexico. Current distribution of this species is unclear because many of the early collection sites have been altered by human activities. In the Grand Canyon, it becomes especially visible during midsummer. Look for it at Clear, Bright Angel, and Stone Creeks, among others.

Fluffgrass *Dasyochloa pulchellum* —Photograph © 2006 by Celia Southwick

Satintail *Imperata brevifolia*
—Photograph © 2006 by Glenn Rink

Satintail *Imperata brevifolia*
—Photograph © 2006 by Kristin Huisinga

Scratchgrass
Muhlenbergia asperifolia

GRASS FAMILY (Poaceae/Gramineae)

Plants: Perennial grass, 10 to 50 centimeters tall. **Leaves:** Narrow, flat to folded, 2 to 8 centimeters long. **Flowers:** In diffuse panicles that are 5 to 25 centimeters long. **Fruit:** Grain. **Flowering Season:** May to October. **Elevation:** 1,400 to 3,100 feet.

When seen from a distance, scratchgrass has a soft, feathery, weblike appearance. Its delicate inflorescence is often half the size of the whole plant. You may see only remnants of it, because the flowering stalks detach when mature, aiding in dispersal.

Without flowers, the narrow leaves and creeping habit of scratchgrass resemble those of Bermuda grass (*Cynodon dactylon*). The ligule of scratchgrass is membranous and lacks the conspicuous, long, white hairs of Bermuda grass. A recent exotic arrival to the Grand Canyon, **witchgrass (*Panicum capillare*)** has similar flowering stalks. You can differentiate it from scratchgrass by its much broader, very hairy leaves and robust habit.

The genus *Muhlenbergia* comprises a large portion of the grass flora in arid and semiarid regions, and our species is well-established across most of North America. In the Grand Canyon, it is widespread along the length of the river and in side canyons, growing on moist ground and dry slopes.

The genus name honors Gotthilf Heinrich E. Muhlenberg (1753–1815), who was a pastor and self-taught botanist. When you walk through a patch of this grass with shorts on, you will realize that *asperifolia* refers to the rough nature of the leaves.

Phragmites, Common Reed
Phragmites australis
(*P. communis*)

GRASS FAMILY (Poaceae/Gramineae)

Plants: Perennial grass, 1 to 4 meters tall. **Leaves:** Alternate, broad, flat, 20 to 45 centimeters long. **Flowers:** In plumelike panicles that are 15 to 50 centimeters long. **Fruit:** Grain. **Flowering Season:** July to October. **Elevation:** 1,200 to 6,400 feet.

Dense stands of phragmites crowd the banks of the Colorado River, resembling fields of corn. The flowering stalks change from grayish purple to straw-colored at maturity, providing a rich array of riverside scenery into the fall. During the winter, the interconnected, vertical stems die down to the base. Each spring, the rhizomes produce new vertical stems that then spread aboveground from stolons and share the river banks and marshes with other wetland plants such as cattails (*Typha* species), bulrushes (*Scirpus* species), sedges (*Carex* species), and rushes (*Juncus* species).

Phragmites may be confused with the exotic bunchgrasses, Ravenna grass (*Saccharum ravennae*) and pampus grass (*Cortaderia* species) because they all have showy, plumelike flowering stalks. However, phragmites is not a bunchgrass and the leaves alternate singly along the culm.

Phragmites may have the widest distribution of any flowering plant—it occurs on every continent and in nearly all of the contiguous forty-eight states. In the eastern United States, it is sometimes considered a noxious weed because it crowds out native species in wetlands, especially those altered by human activities. In the Grand Canyon, it grows in soils with fine particles, like those found in backwaters, the entire length of the river and in other scattered locations.

Phragmites is a significant element in creation stories and ceremonies of several southwestern Native American tribes, who also use the stems and roots as medicine. Mats, roofing, pipes, arrow shafts, and weaving rods can be fashioned from the hollow stems and leaves.

Scratchgrass *Muhlenbergia asperifolia*
—Illustration © 2006 by Kate Watters

Witchgrass *Panicum capillare*
—Illustration © 2006 by Kate Watters

Phragmites *Phragmites australis* —Photograph © 2006 by Kate Watters

Galleta
GRASS FAMILY (Poaceae/Gramineae)

Pleuraphis jamesii
(Hilaria jamesii)

Plants: Perennial bunchgrass, 15 to 50 centimeters tall. **Leaves:** Narrow, flat or folded, 1 to 16 centimeters long. **Flowers:** In racemose spikes that are 2 to 9 centimeters long. **Fruit:** Grain. **Flowering Season:** May to September. **Elevation:** 1,800 to 5,500 feet.

Galleta, a desert grassland species, often grows in dense stands, catching the glow of the late afternoon sun. Even when the flowers have matured and fallen, the characteristic zig-zag of the flowering stalks indicates that you have found galleta.

Two *Pleuraphis* species in the Grand Canyon resemble each other and have slightly overlapping ranges. **Big galleta (*P. rigida*)** can grow nearly twice the size of galleta and has tightly curled, densely woolly hairs in the upper portion of the sheaths and around the ligule. Galleta has long straight hairs at the sheath summit and nodes and a membranous ligule.

Galleta occurs from Wyoming south to Texas and west to California. River travelers are likely to encounter galleta first because it is known from Nankoweap Canyon to Havasu Creek. Big galleta prefers warmer temperatures and is present from Waltenberg Canyon downstream to Grand Wash Cliffs. The Grand Canyon is one of the places where these species co-mingle at the periphery of their ranges.

Both species are important forage plants. Hopi and Navajo basket weavers use the culms of galleta as fill in coil baskets. Big galleta often grows with cholla cacti (*Cylindropuntia* species), protecting the cholla seedlings, while in turn, the young cholla prevent animals from grazing the grass.

Rabbitfoot Grass
GRASS FAMILY (Poaceae/Gramineae)

Polypogon monspeliensis

Plants: Exotic annual grass, 3 to 80 centimeters tall. **Leaves:** Narrow, flat, lax, 1 to 21 centimeters long. **Flowers:** In cylindrical, densely compact panicles that are 1 to 15 centimeters long. **Fruit:** Grain. **Flowering Season:** March to October. **Elevation:** 1,200 to 7,700 feet.

As its common name suggests, the soft-hairy, spikelike panicles of rabbitfoot grass bear a striking resemblance to a lucky rabbit foot you might find on a key chain. Although an annual, its stems fall over at the base, rooting at the nodes and giving it the appearance of a rhizomatous perennial grass.

Rabbitfoot grass is native to Eurasia and Africa and is now widespread throughout the United States. In the Grand Canyon, it grows in side canyons and along the banks the entire length of the river. It prefers habitats with moist soils, extending to high elevations on the North Rim.

The fluffy flowering stalks of rabbitfoot grass are similar to those of satintail (*Imperata brevifolia*), which is a much taller, erect perennial with a longer, more slender panicle. There are two other *Polypogon* species in the Grand Canyon. A less common, perennial relative, **ditch polypogon (*P. interruptus*)** is similar in appearance but looks as if it has lost clumps from its flowering stalk, hence its specific epithet, *interruptus*. **Waterbent grass (*P. viridis*)** shares similar habitats and is as abundant as rabbitfoot grass, but this species lacks the long awns that give rabbitfoot grass its fluffy appearance.

The word *poly* means "much" and *pogon* means "bearded," referring to the exceedingly hairy spikelets.

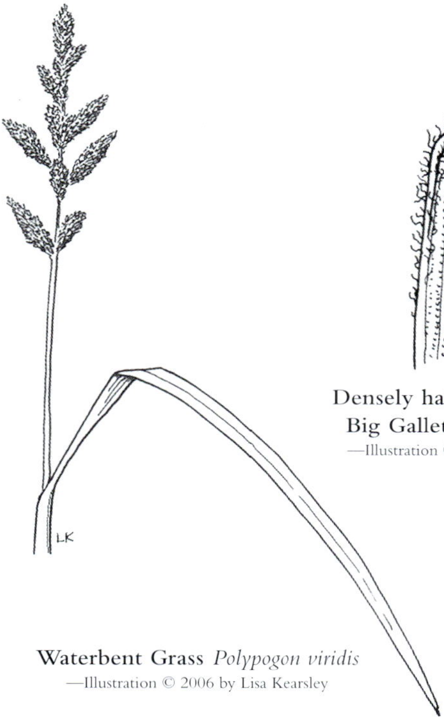

Waterbent Grass *Polypogon viridis*
—Illustration © 2006 by Lisa Kearsley

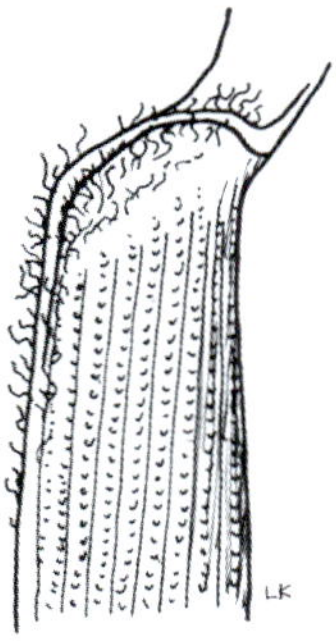

Densely hairy leaf sheaths of Big Galleta *Pleuraphis rigida*
—Illustration © 2006 by Lisa Kearsley

Galleta *Pleuraphis jamesii*
—Illustration © 2006 by Lisa Kearsley

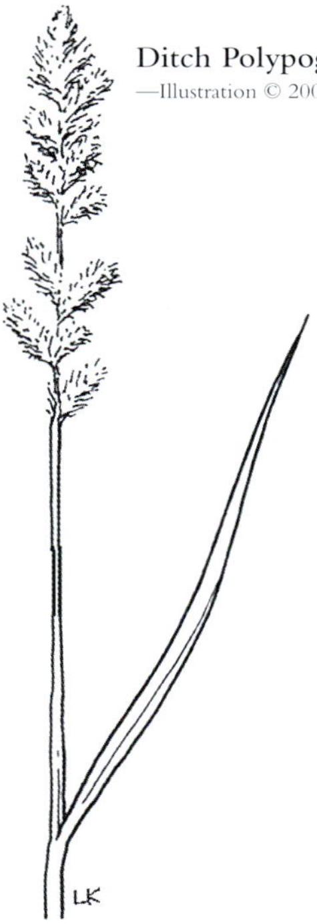

Ditch Polypogon *Polypogon interruptus*
—Illustration © 2006 by Lisa Kearsley

Rabbitfoot Grass *Polypogon monspeliensis* —Photograph © 2006 by Kate Watters

Ravenna Grass
GRASS FAMILY (Poaceae/Gramineae)

Saccharum ravennae
(*Erianthus ravennae*)

Plants: Exotic perennial bunchgrass, 2 to 4 meters tall. **Leaves:** Flat or folded, firm, rough-margined, to 1.5 meters long. **Flowers:** In plumelike panicles that are 25 to 60 centimeters long. **Fruit:** Grain. **Flowering Season:** June to October. **Elevation:** 1,300 to 3,500 feet.

Other riparian plants often camouflage Ravenna grass until its enormous, silvery plumes are tall enough to sway in the breeze. One mature plant can have up to one hundred flowering stalks that produce several thousand seeds, making it an aggressive and successful invader. Wind and water readily disperse the tiny, fluffy seeds, which are only viable for a short period.

Ravenna grass, a relative of sugarcane, is native to Europe, but now grows in at least fifteen states. It was planted as an ornamental in the Glen Canyon region and spread rapidly downstream. Now it grows on beaches, slopes, marshy areas, and gravel bars along the length of the Grand Canyon, but has yet to spread into any side canyons. A successful manual-removal control program began in the early 1990s and continues to keep this invader at bay. The genus name *Ravenna* may be named for the town of Ravenna, Italy, where it flourishes.

Pampus grass (*Cortaderia* species), another exotic bunchgrass, arrived in the Grand Canyon area in the last few decades. Overall, it is more robust and the silvery white seedheads are stouter and more dense. Phragmites (*Phragmites australis*) has similar plumelike flowering stalks but it lacks the bunchlike habit of Ravenna grass and has alternating leaf blades along the length of its culm. To confirm Ravenna grass, look for the V-shaped leaf blades that display a prominent white midvein and sheaths with dense hairs at the base.

Top: **Ravenna Grass** *Saccharum ravennae*
—Photograph by Kate Watters/National Park Service

Bottom Left: **Pampus Grass**
Cortaderia species
—Photograph by Kate Watters/National Park Service

Bottom Right: **Ravenna Grass inflorescence** *Saccharum ravennae*
—Photograph © 2006 by Dan Hall

Sand Dropseed
GRASS FAMILY (Poaceae/Gramineae)

Sporobolus cryptandrus

Plants: Perennial bunchgrass, 30 to 100 centimeters tall. **Leaves:** Flat to involute, 10 to 30 centimeters long, with hairy ligules and sheaths. **Flowers:** In open panicles that are 10 to 40 centimeters long. **Fruit:** Grain. **Flowering Season:** April to October. **Elevation:** 1,200 to 8,000 feet.

Nearly one hundred dropseed species (*Sporobolus* species) occur worldwide. One of the most common grass genera in the Southwest, more than fifteen dropseed species inhabit Arizona. Six species of dropseed live in the Grand Canyon, all of which bear long tufts of hairs on their ligules.

Sand dropseed is our most common species. Its sheaths enclose the lower part of the opening flowering stalk, making it appear stuck with its arms in a straight jacket. A similar species, **mesa dropseed (*S. flexuosus*)** has a more fully open, drooping flowering stalk with tangled branches. It also displays conspicuous tufts of spiny hairs in the axils just above the panicle branches that people remember as "hairy shoulders." **Alkali sacaton (*S. airoides*)** forms larger, denser clumps and bears gigantic, open-flowering stalks.

Two other perennial dropseeds are quite different in that their spikelike, tightly contracted panicles never spread open. The culms of **giant dropseed (*S. giganteus*)** often reach heights of 2 meters and bear thicker, longer panicles up to 70 centimeters long. **Spike dropseed (*S. contractus*)** is more petite in all respects than its giant brother, including its more slender, shorter panicles that extend from 15 to 50 centimeters long.

In the Grand Canyon, sand dropseed grows on dry, open, sandy dunes, beaches, and rocky slopes the length of the river and extending to the rims. The other dropseed species mentioned here are found in similar habitats the length of the river but restricted to lower elevations.

Dropseeds were a staple, late-season food of indigenous people who actively harvested the tiny, numerous seeds by shaking them into baskets, pottery, or blankets. Wildlife also treasure these grasses and their digestive juices may enhance germination of those seeds that persist in droppings.

Long tufts of hairs on ligules
of all *Sporobolus* species
—Illustration © 2006 by Lisa Kearsley

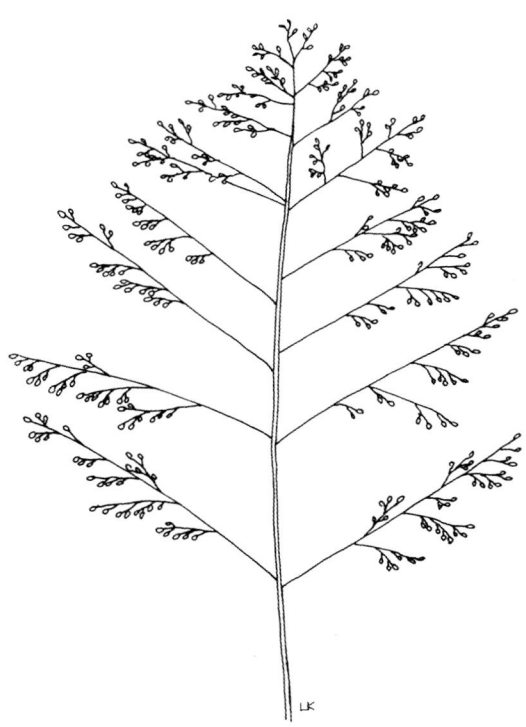

Alkali Sacaton *Sporobolus airoides*
—Illustration © 2006 by Lisa Kearsley

Sand Dropseed *Sporobolus cryptandrus*
—Illustration © 2006 by Lisa Kearsley

Mesa Dropseed
Sporobolus flexuosus
—Illustration © 2006 by Lisa Kearsley

Spikelike panicles of Giant Dropseed (*Sporobolus giganteus*) **and Spike Dropseed** (*Sporobolus contractus*)
—Illustration © 2006 by Lisa Kearsley

Six-Weeks Fescue
GRASS FAMILY (Poaceae/Gramineae)

Vulpia octoflora
(*Festuca octoflora*)

Plants: Annual grass, 3 to 30 centimeters tall. **Leaves:** Narrow, involute, short, with long, hairy ligules. **Flowers:** In narrow panicles that are 1 to 10 centimeters long. **Fruit:** Grain. **Flowering Season:** April to June. **Elevation:** 1,200 to 6,800 feet.

Six-weeks fescue packs germination, growth, flowering, and seed dispersal into six short weeks in late spring. Repeat photography in the Grand Canyon has demonstrated that several perennial grass species can live for more than a century, but six-weeks fescue has a different strategy: it has reduced its size and lifespan so that all of its resources can go into seed production. Six-weeks grama (*Bouteloua barbata*) has a similar strategy and stature, but it flowers in the late summer. These separate flowering seasons make them easy to tell apart.

Six-weeks fescue looks somewhat like the exotic annual **Mediterranean grasses (*Schismus* species)**, which have aggressively spread into many desert areas of the Southwest. Look for the pointed awns on six-weeks fescue, which are lacking in the Mediterranean grasses.

Six-weeks fescue has an extensive range in lower elevations throughout North America and it thrives in disturbed areas. In the Grand Canyon, look for it on open slopes along the river in a wide variety of habitats throughout the Inner Canyon, extending to the North and South Rims.

Rush
RUSH FAMILY (Juncaceae)

Juncus species

Though closely related to sedges, rushes are in their own family because each flower has six sepals and petals compared to sedge family (Cyperaceae) flowers that have one bract at the base of each flower. The rounded stems of rushes arise from rhizomes and have unique hard subdivisions, called *crosswalls*, which can be felt when gently sliding your fingers up the stem.

Along the river, rushes are more widespread than sedges, and three common species are described here. The compact, rounded, chestnut brown inflorescence of **Torrey's rush (*Juncus torreyi*)** make it easy to identify. Unlike Torrey's rush, the more openly spaced inflorescences of **jointed rush (*J. articulatus*)** and **wire rush (*J. balticus*)** bear flowers clustered in whorls on branchlet ends. Look for the more developed leaves at the base and along the stem in jointed rush. Compare these to the reduced, nearly lacking leaves that are restricted to the very base of the stems in wire rush.

Eight of two hundred species of rushes found worldwide occur in the Grand Canyon. Look for them in the moist soil along the river's edge and in wet side canyons. The stems of many rush species are used to make baskets, with hundreds of stems often required for each basket.

Six-Weeks Fescue *Vulpia octoflora*
—Illustration © 2006 by Bronze Black

Mediterranean Grasses *Schismus* species
—Illustration © 2006 by Bronze Black

Left: Jointed Rush showing well-developed leaves
Juncus articulatus —Illustration © 2006 by Lisa Kearsley

Right: Wire Rush showing reduced lower leaves
Juncus balticus —Illustration © 2006 by Lisa Kearsley

Torrey's Rush *Juncus torreyi*
—Photograph © 2006 by Kate Watters

SEDGE FAMILY

The riparian belt bordering the river and side canyon waterways is a unique mix of rather inconspicuous herbs, grasses, and other semiaquatic species. Lacking showy flowers, plants in the sedge (Cyperaceae) and rush (Juncaceae) families are often overlooked or confused with grasses. A familiar saying helps identify them: "Sedges have edges, rushes are round, grasses are hollow from their knees to the ground."

Worldwide there are over four thousand species in the sedge family, mostly occupying damp or marshy areas. People have used them for centuries as material for baskets, mats, hats, and skirts, and as food and perfume. For purposes of simplification, we are presenting the most common genera in the Grand Canyon (*Carex* species, *Cladium* species, *Cyperus* species, *Eleocharis* species, *Schoenoplectus* species).

Sedge *Carex* species
SEDGE FAMILY (Cyperaceae)

At least six of the Grand Canyon's twenty sedge species are found along the Colorado River, and the most common species is **leafy sedge (*Carex aquatilis*)**. It forms large, vibrant green tufts along the shoreline, especially in Marble Canyon, and upon closer examination, you will notice its long, V-shaped, dark green leaves. Nesting birds and small mammals eat young shoots and use dense stands of sedge species for cover.

Sawgrass *Cladium californicum*
SEDGE FAMILY (Cyperaceae)

Plants: Rhizomatous perennial herb, to 2 meters tall. **Leaves:** Coarse, flat, saw-tooth margined, to 1.5 centimeters wide. **Flowers:** In panicle-like clusters that are up to 50 centimeters long. **Fruit:** Smooth achene. **Flowering Season:** April to October. **Elevation:** 1,300 to 3,800 feet.

Sawgrass is the most conspicuous member of the sedge family in the Grand Canyon because of its lofty, drooping, rusty brown flowering stalks and saw-toothed leaf margins. From a distance, you might confuse this with beargrass (*Nolina microcarpa*), which has a comparable stature, sharp-edged leaves and sometimes grows near sawgrass. However, beargrass resides in its own unique family, the nolina family, and thus has a distinctly different flower structure. The shaggy, rusty brown flowering stalks of sawgrass distinguish it from beargrass, which has clusters of creamy white flowers on tall stalks. Sawgrass, a water-loving plant, is a conspicuous component of the vegetation in many side canyons.

Sawgrass is restricted to marshes and saline seeps in California, Nevada, New Mexico, Utah, and Arizona, and south into Central America. While considered a botanical relict from an earlier, warmer period, it finds refuge in the well-developed riparian areas of the Grand Canyon. If you are traveling on the river, start looking for it around River Mile 65 and farther downstream where it is most common in side canyons at streams, seeps, and springs.

Flat Sedge *Cyperus* species
SEDGE FAMILY (Cyperaceae)

There are at least two species of flat sedge in the Grand Canyon, both with restricted distributions. *Cyperus* species have long, leaflike bracts enclosing from below several headlike clusters that are borne at the tops of the stems. Egyptians cultivated one species of flat sedge to make papyrus, an ancient paper, from overlaid strips of the inner stems.

Flat Sedge *Cyperus erythrorhizos*
—Illustration © 2006 by Lisa Kearsley

Sawgrass *Cladium californicum*
—Photograph © 2006 by Raechel Running RMRfotoarts.com

Sedge *Carex aquatilis*
—Illustration © 2006 by Lisa Kearsley

Flowers of Sawgrass *Cladium Californicum*
—Illustration © 2006 by Glenn Rink

Spikerush
Eleocharis species
SEDGE FAMILY (Cyperaceae)

In the Grand Canyon, at least fhree of th six species of spikerush occur along the rier and in side canyons. The solitary, conelike spikelets are borne at the tips of the slender flowering stalks that lack the leafy bracts of *Cyperus* species. Chinese water chestnuts are the corms of one spikerush species. The genus name, *Eleocharis*, is from the Greek words *helos*, meaning "marsh" and *charis*, meaning "grace," referring to the slender habit of these petite plants and their preference for marshes.

Bulrush, Tule
Schoenoplectus species (*Scirpus* species)
SEDGE FAMILY (Cyperaceae)

Six species of bulrush prosper in marshes along the river and in wet side canyons, such as the Little Colorado and Havasu. The subterminal spikelet clusters and large round stems distinguish bulrushes from other genera in the sedge family, many of which have terminal spikes. Whole ecological communities rely on diverse marsh habitats where bulrushes flourish. The pliable stems offer material for baskets, mats, and boats. The rhizomes provide food, and wildlife find cover in the tall stems.

Spikerush *Eleocharis parishii*
—Photograph © 2006 by Kristin Huisinga

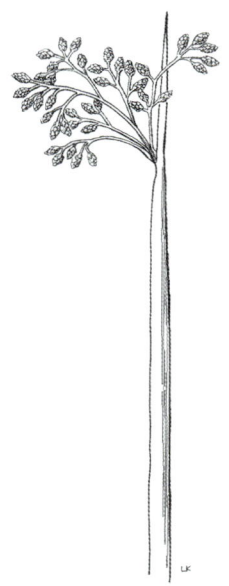

Bulrush *Schoenoplectus* species
—Illustration © 2006 by Lisa Kearsley

TREES

Desert-Willow *Chilopsis linearis*
BIGNONIA FAMILY (Bignoniaceae)

Plants: Deciduous tree, to 7 meters tall. **Leaves:** Linear, 10 to 26 centimeters long. **Flowers:** Terminal, light pink to lavender, 2 to 5 centimeters long. **Fruit:** Linear pod. **Flowering Season:** May to August. **Elevation:** 1,200 to 3,000 feet.

The voluptuous, two-lipped flowers of desert-willow distinguish it from true willows (*Salix* species), which have rather inconspicuous flowers. Although part of a large tropical family with few relatives in the Southwest, desert-willow thrives in dry washes, sending down roots nearly twice its height in search of water. A waxy protective layer coats the leaves, helping to retain moisture during dry periods.

Desert-willow grows throughout the Southwest and into northern Mexico. It is uncommon in the Grand Canyon, known primarily from washes and canyons, such as Havasu Canyon and 209 Mile Wash, and along the upper part of Diamond Creek Road on the Hualapai Reservation.

Carpenter bees (*Xylocopa* species) are the flowers' primary pollinators. The genus name, *Chilopsis*, is from the Greek *cheilos* and *opsis*, meaning "resembles lips."

Desert-Willow *Chilopsis linearis*
—Photograph © 2006 by Kathy Darrow

Western Hop Hornbeam
Ostrya knowltonii
BIRCH FAMILY (Betulaceae)

Plants: Deciduous tree, to 10 meters tall. **Leaves:** Ovate, doubly toothed, to 8 centimeters long. **Flowers:** In catkins, inconspicuous. **Fruit:** Winged nutlet. **Flowering Season:** April to May. **Elevation:** 4,000 to 8,900 feet.

Western hop hornbeam has scaly, ashy gray bark and dangling fruit clusters that resemble the fruit of hops. Like other members of the birch family, it has finely toothed leaves with conspicuous veins.

Western hop hornbeam is likely a relict species from colder and moister times and is now restricted to alcoves, north-facing slopes, and riparian areas in Utah, Arizona, New Mexico, and Texas. In the Grand Canyon, it grows in side canyons and otherwise moist and shady sites scattered throughout the Inner Canyon.

The genus name, *Ostrya*, is Greek for "hardwood tree," in reference to the dense wood common to all hop hornbeams. This species is named for Frank Knowlton (1860-1926), a botanist who found the tree growing just below the rim of the Grand Canyon.

Crucifixion-Thorn
Canotia holacantha
BITTERSWEET FAMILY (Celastraceae)

Plants: Thorny shrub to small tree, to 6 meters tall. **Leaves:** Reduced to scales. **Flowers:** In racemes, green to white, five-petaled. **Fruit:** Capsule. **Flowering Season:** May to August. **Elevation:** 2,000 to 4,500 feet.

While not abundant in the Grand Canyon, this small tree with sturdy thorns will pique your curiosity. Crucifixion-thorn relies upon its green, photosynthetic stems and branches to produce food because its leaves are reduced to meer scales. While the flowers are inconspicuous, they house a large nectar disk. Upon maturity, they produce reddish brown, egg-shaped capsules that split, releasing their seeds from the top.

Crucifixion-thorn only grows in Arizona and Mexico. It covers large expanses in the Sonoran and Mojave Deserts, where it is sometimes confused with foothill paloverde (*Cercidium microphyllum*). Crucifixion-thorn is more frost tolerant, allowing it to extend as far north as the western Grand Canyon region, where it is most conspicuous along Diamond Creek Road.

Parts of this shrub are used in teas to treat symptoms of giardia, and the wood is used for firewood.

Western Hop Hornbeam *Ostrya knowltonii* —Illustration © 2006 by Mar-Elise Hill

Crucifixion-Thorn *Canotia holacantha* —Photograph © 2006 by Kristin Huisinga

Birchleaf Buckthorn
BUCKTHORN FAMILY (Rhamnaceae)

Frangula betulifolia ssp. *obovata*
(*Rhamnus betulifolia* var. *obovata*)

Plants: Deciduous shrub or small tree, to 2.5 meters tall. **Leaves:** Alternate, elliptic to oblong, finely toothed, 2 to 15 centimeters long. **Flowers:** In axillary clusters, green to brown, inconspicuous. **Fruit:** Berry. **Flowering Season:** April to June. **Elevation:** 1,700 to 5,900 feet.

Birchleaf buckthorn often intermingles with other deciduous plants. It has unusually large, dark green leaves with waxy, distinctly veined upper surfaces and undersides that are finely hairy and coarse to the touch. The dark purple to black berries ripen in the fall.

Birchleaf buckthorn inhabits Nevada, Utah, Arizona, New Mexico, and Texas. In the Grand Canyon, it grows in moist, shady side canyons and hanging gardens from Saddle Canyon to River Mile 170. If you have a keen eye, you will spot several of these shrubs from the river between Matkatamiba and Havasu Canyons.

Although wildlife and birds relish the dry berries, they are not palatable for humans. Buckthorn bark treats problems such as constipation, rheumatism, and inflammation.

Pale Hoptree
CITRUS FAMILY (Rutaceae)

Ptelea trifoliata ssp. *pallida*

Plants: Deciduous shrub or small tree, to 7 meters tall. **Leaves:** Palmately compound into three lanceolate to ovate leaflets, each leaflet 14 centimeters long. **Flowers:** In clusters that are 2 to 5 centimeters wide; small greenish white flowers, with four petals. **Fruit:** Round, winged samara. **Flowering Season:** April to June. **Elevation:** 2,000 to 8,200 feet.

Pale hoptree, a multistemmed tree, has shiny, compound leaves with sunken glands that look like hundreds of tiny holes when held up to light. These glands are filled with terpenes, oily compounds, that give off a citrus odor when crushed, revealing its citrus family lineage.

In the United States, pale hoptree occurs from Utah and Arizona east to Texas. In the Grand Canyon, it grows in warm microhabitats on the North and South Rims and in moist side canyons the length of the canyon, although it is not known from sites along the Colorado River. Look for it in North and Saddle Canyons and along the Kaibab Trail.

Historically, the leaves were made into a poison for arrow tips in hunting or warfare. The genus name, *Ptelea*, is Greek for "elm," a reference to the winged fruit. The specific epithet, *trifoliata*, refers to each leaf having three leaflets.

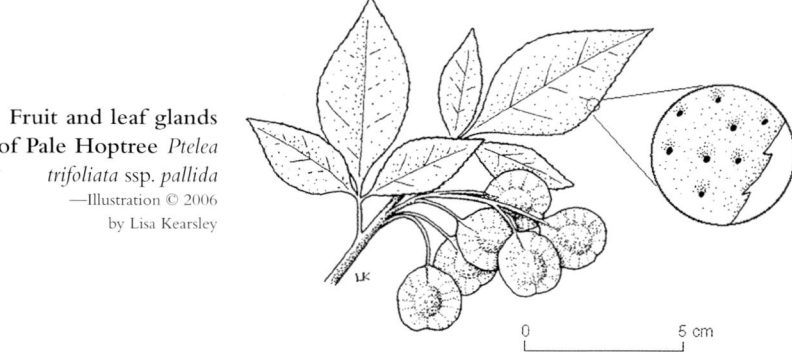

Fruit and leaf glands of Pale Hoptree *Ptelea trifoliata* ssp. *pallida*
—Illustration © 2006 by Lisa Kearsley

Birchleaf Buckthorn *Frangula betulifolia* ssp. *obovata* —Photograph © 2006 by David Edwards

Pale Hoptree *Ptelea trifoliata* ssp. *pallida* —Photograph © 2006 by Kate Watters

One-Seed Juniper
CYPRESS FAMILY (Cupressaceae)

Juniperus monosperma
(*J. occidentalis* var. *gymnocarpa*)

Plants: Multibranched, evergreen tree, to 8 meters tall, with shreddy bark. **Leaves:** Scalelike, green to dark green, to 3 millimeters long. **Flowers:** Male and female cones on separate trees. **Fruit:** Berry-like cone. **Flowering Season:** March to April. **Elevation:** 1,200 to 5,500 feet.

Only a few scattered individuals of one-seed juniper grow along the river corridor, and one tree in Marble Canyon bears the mark of history, with the initials of Harry McDonald from Robert Brewster Stanton's 1889 expedition. Juniper and other species in the cypress family are relicts of the wetter and cooler climate during Pleistocene time. One-seed juniper, also called *cedar*, resembles red cedar (*Juniperus virginiana*), the most common juniper species in the United States.

Members of the cypress family are the most widely distributed of the gymnosperms and occur on every continent except Antarctica. One-seed juniper is found in Texas, New Mexico, Colorado, and Arizona, and is rare in Utah and Oklahoma. At lower elevations in the Grand Canyon, it grows singly or in stands. As you journey from low deserts to higher elevations in the western states, piñon pine (*Pinus edulis*) and **Utah juniper (*J. osteosperma*)** often replace one-seed juniper. Where the species overlap, you can distinguish Utah juniper from the many-branched one-seed juniper by its single, definite trunk and larger fruit.

All juniper species played an integral role in the survival of native people throughout the West, providing fuel, building material, medicine, food, and a variety of ceremonial items. Birds and coyotes relish the tasty reddish blue fruit and then deposit the seeds in their droppings, helping disperse this species. The higher-elevation **common juniper (*J. communis*)** provides some of the flavor in gin.

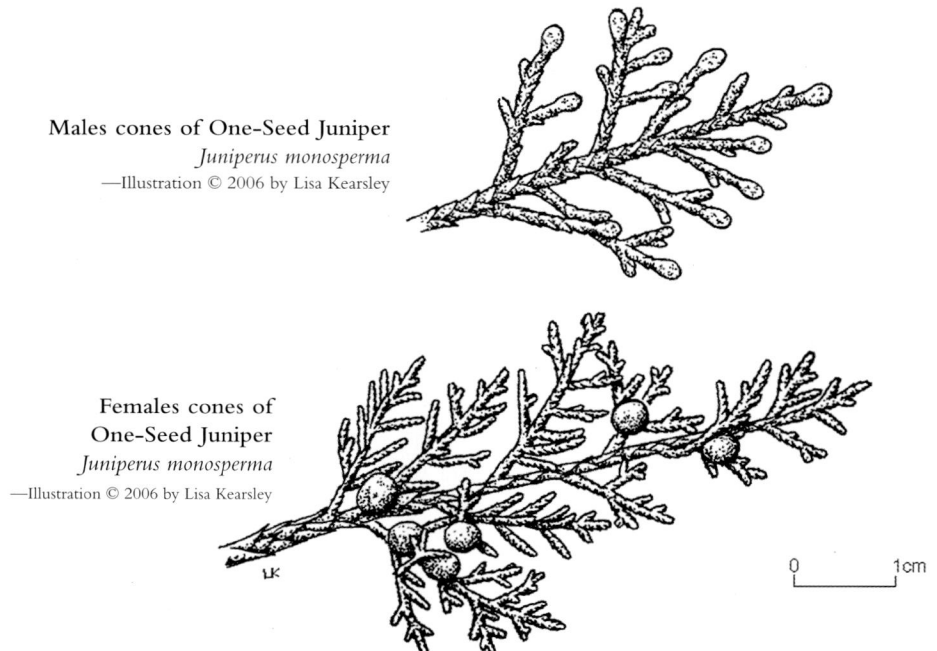

Males cones of One-Seed Juniper
Juniperus monosperma
—Illustration © 2006 by Lisa Kearsley

Females cones of One-Seed Juniper
Juniperus monosperma
—Illustration © 2006 by Lisa Kearsley

1889 Harry McDonald inscription in trunk of a One-Seed Juniper (*Juniperus monosperma*) in Marble Canyon —Photograph © 2006 by Kristin Huisinga

Utah Juniper
Juniperus osteosperma
—Photograph © 2006 by Kristin Huisinga

Netleaf Hackberry
ELM FAMILY (Ulmaceae)

Celtis laevigata var. *reticulata*
(*Celtis reticulata*)

Plants: Deciduous tree, to 9 meters tall. **Leaves:** Alternate, triangular to ovate, finely saw-toothed margins, long-pointed tips, 2 to 8 centimeters long. **Flowers:** In axils, solitary or clustered, inconspicuous. **Fruit:** Orange to dark red drupe. **Flowering Season:** March to April. **Elevation:** 1,500 to 6,000 feet.

The spreading crown of netleaf hackberry tops a stout, gnarled trunk. The bright green leaves turn yellow in the fall, and clusters of insect galls often give them a warty appearance. The roughly textured leaves are asymmetrical at the base. Often living well over one hundred years, netleaf hackberry typically grows in desert riparian zones in the Southwest. The oldest known specimens in the Grand Canyon germinated in 1869, the year that John Wesley Powell made his pioneering journey down the Colorado River.

Netleaf hackberry occurs from Washington and Idaho, south to northern Mexico, and east to Kansas and Nebraska. In the Grand Canyon, these trees grow at scattered locations throughout the Inner Canyon in canyon bottoms, along streams, and on slopes just above the pre-dam high-water line.

The caterpillars of hackberry emperor butterflies (*Asterocampa* species) eat the leaves of netleaf hackberry, and the butterflies are one of the trees' primary pollinators. Netleaf hackberry fruit offers an important food source for birds, small mammals, and people. To aid digestion, native people boiled or pounded the fruit into a pulp, seeds and all, and then mixed it with dried cornmeal or animal fat.

Arizona Boxelder
MAPLE FAMILY (Aceraceae)

Acer negundo var. *arizonicum*
(*Acer negundo* ssp. *californicum*)

Plants: Deciduous tree, to 12 meters tall. **Leaves:** Opposite, pinnately compound, with three to five leaflets, 2 to 10 centimeters long. **Flowers:** Male flowers in bundles; female flowers in drooping racemes that are up to 4 centimeters. **Fruit:** Paired, winged samaras. **Flowering Season:** March to May. **Elevation:** 2,400 to 7,900 feet.

Arizona boxelder, a large, broadly crowned tree, has lush greenery that stands out in the dry, desert landscape of the Grand Canyon. Its coarsely toothed leaves and winged fruits, which resemble those of other members of the maple family, distinguish it from other trees in our region. It is unusual among American maples in that the male and female flowers are on separate trees and that its leaves are compound.

Arizona boxelder resembles another deciduous tree, velvet ash (*Fraxinus velutina*), which has leaves divided into five to seven leaflets and clusters of many winged samaras. While not abundant in the Grand Canyon, poison-ivy (*Toxicodendron rydbergii*) also has leaves with three leaflets, but differs in that the leaves are alternate along the stem.

While boxelder grows throughout North America, our variety is restricted to Arizona and New Mexico. In the Grand Canyon, it is present on both rims, in side canyons with permanent water, and occasionally scattered along the river.

Boxelder is a fast-growing, relatively short-lived, and weak-stemmed tree, which wind, floods, heart rot, and insects readily injure. The flowers are primarily wind pollinated, although bees may also visit them. Deer browse young plants, and birds and small mammals eat the winged fruits, especially during the winter. Native Americans used the inner bark, sap, and branches for food, medicinal teas, candy, and charcoal paint. The sap is high in sugar and can be used to make syrup, much like sap from maples in the eastern United States.

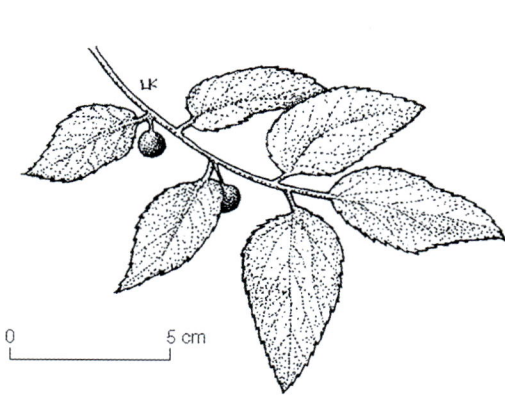

Asymmetrical leaf bases of Netleaf Hackberry
Celtis laevigata var. *reticulata*
—Illustration © 2006 by Lisa Kearsley

Netleaf Hackberry *Celtis laevigata* var. *reticulata*
—Photograph © 2006 by Kristin Huisinga

Arizona Boxelder *Acer negundo* var. *arizonicum*
—Photograph © 2006 by Kristin Huisinga

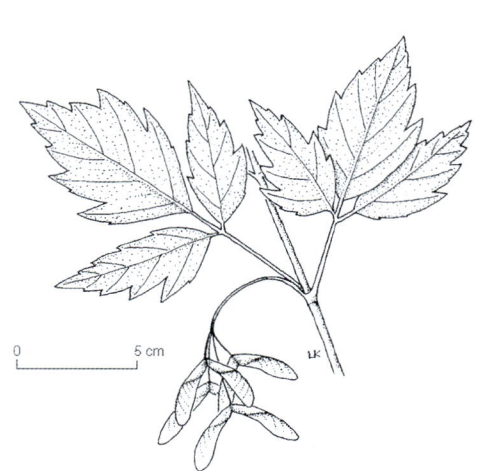

Fruit and leaves of Arizona Boxelder
Acer negundo var. *arizonicum*
—Illustration © 2006 by Lisa Kearsley

Shrub Live Oak   *Quercus turbinella*
OAK FAMILY (Fagaceae)

Plants: Evergreen shrub or small tree, 1 to 5 meters tall. **Leaves:** Alternate, oblong to elliptic, grayish to dark green, spiny-margined, 1 to 4.5 centimeters long. **Flowers:** Solitary or in catkins; males drooping, short; females spikelike; flowers inconspicuous. **Fruit:** Acorn. **Flowering Season:** April to September. **Elevation:** 2,200 to 8,000 feet.

Shrub live oak is smaller than the majestic oaks found in many areas in the United States. Look for empty acorn cups on the branches of this evergreen oak, especially when the squirrels are hungry. A common resident of Arizona's chaparral zone, it forms thickets in association with other shrubs and trees, sometimes turning a pleasant walk into an arduous bushwhacking adventure.

Shrub live oak flourishes throughout canyon country at lower elevations from Texas to California and south to Mexico. In the Grand Canyon, you can find it in many side canyons including Kwagunt, Nankoweap, and Hermit Canyons.

The very spiny, hollylike, evergreen leaves distinguish shrub live oak from the deeply lobed, deciduous leaves of **Gambel oak (*Quercus gambellii*),** which is also found in side canyons. Red barberry (*Mahonia haematocarpa*), another similar shrub, has bluish green, spiny leaves that are divided into three to seven leaflets.

Many indigenous groups throughout the West arranged special camps and trading routes to facilitate the processing and storage of acorns, which were eaten raw or ground into a flour. Most acorns require soaking to remove the bitter tannins. People often gained community status by developing great skill in acorn preparation.

Velvet Ash   *Fraxinus velutina*
OLIVE FAMILY (Oleaceae)   (*F. pennsylvanica* ssp. *velutina*)

Plants: Deciduous tree, 12 to 18 meters tall, bark furrowed with age. **Leaves:** Opposite, pinnately compound, 10 to 25 centimeters long. **Flowers:** In axillary clusters; male and female separate. **Fruit:** Clustered samaras. **Flowering Season:** April to September. **Elevation:** 1,200 to 3,100 feet.

These gently arching lush giants are remnants of a once-widespread deciduous forest. As climates became hotter and drier beginning about 14,000 years ago in late Pleistocene time, plants lacking desert adaptations found refuge in continually wet areas, like the side creeks of the Grand Canyon.

As the most widely distributed of the American ashes, velvet ash occurs throughout North America. In the Grand Canyon, it grows along streams and in side canyons from River Mile 154 to Grand Wash Cliffs, and is particularly abundant at Havasu Creek. Be on the lookout for quirky trees that have managed to establish along the Colorado River.

Another species, **single-leaf ash (*Fraxinus anomala*)** is the only ash that has single, rounded leaves compared to the five to nine leaflets of most ashes. It reaches 7 meters tall, about half the size of velvet ash, and its samaras grow singly. Limited to the southwestern United States, single-leaf ash occurs up to 6,500 feet in the Grand Canyon in the lower reaches of the piñon-juniper woodland and in narrow side canyons and shady alcoves.

With its large stature and tender foliage, velvet ash provides food and shelter for many animals. Native people and European settlers used virtually every part of the velvet ash, making dye from the roots and bark, bows and arrows from the branches, a variety of medicines from the inner bark, an aphrodisiac from the seeds, and prayersticks for ceremony.

Gambel Oak *Quercus gambelii*
—Illustration © 2006 by Bronze Black

Shrub Live Oak *Quercus turbinella*
—Photograph © 2006 by Lisa A. Hahn

Velvet Ash *Fraxinus velutina*
—Illustration © 2006 by Lisa Kearsley

Velvet Ash *Fraxinus velutina*
—Photograph © 2006 by Geoff Gourley

Single-Leaf Ash
Fraxinus anomala
—Photograph © 2006
by Kristin Huisinga

Catclaw Acacia
Acacia greggii
PEA FAMILY (Fabaceae/Leguminosae)

Plants: Deciduous tree, to 5 meters tall. **Leaves:** Bipinnate, with two to seven pairs of leaflets, 2 to 5 centimeters long. **Flowers:** In cylindrical, dangling spikes that are 3 to 4 centimeters long; cream to yellow flowers. **Fruit:** Flattened pod. **Flowering Season:** April to June. **Elevation:** 1,200 to 4,800 feet.

The upright, spare stature of catclaw acacia helps to identify this tree from afar. The lusciously fragrant blossoms fill the air with a delightful aroma of spring. You'll notice bees buzzing around your head as you move closer into the canopy of hanging flower clusters. Catclaw acacia is commonly confused with western honey mesquite (*Prosopis glandulosa* var. *torreyana*), with which it often grows, but in all growing seasons, there are ways to tell them apart. Catclaw acacia has small, curved prickles, not straight, paired spines like those of mesquite. The leaves of catclaw acacia are altogether smaller than the also bipinnately compound leaves of mesquite. The chestnut brown pods of catclaw are constricted between the seeds unlike the longer, straight pods of mesquite. From the river, they may be hard to tell apart but mesquite is less abundant from River Mile 76 to National Canyon while catclaw remains a common sight on the steep slopes throughout the Inner Gorge, the steep-walled, schist-dominated area downstream of River Mile 76.

The genus Acacia is distributed worldwide throughout the tropics and subtropics, especially in arid areas. Catclaw acacia occurs from California to Texas, and south to Mexico. In the Grand Canyon, it grows throughout the Inner Canyon, first appearing around River Mile 39 and continuing the length of the river, where it is commonly a component of the community of plants that grow along the pre-dam high-water line.

Although less desirable than mesquite pods, the pods of catclaw acacia are a food for some Native American groups, and the young branches can be fashioned into cradleboards and other small objects. The epithet, *greggii*, celebrates Josiah Gregg (1806–1850), a frontier author and plant collector in the Southwest.

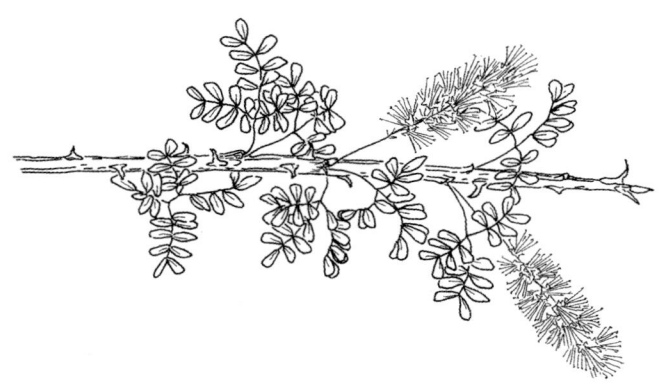

Flowering branch of Catclaw Acacia *Acacia greggii*
—Illustration © 2006 by Mar-Elise Hill

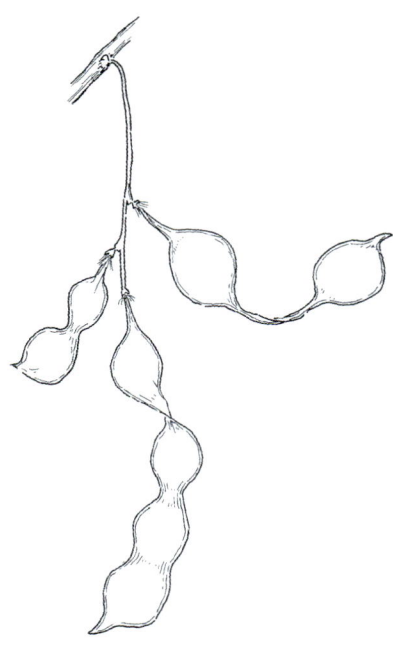

Fruit of Catclaw Acacia *Acacia greggii*
—Illustration © 2006 by Mar-Elise Hill

Catclaw Acacia *Acacia greggii*
—Photograph © 2006 by Lisa A. Hahn

Catclaw Acacia *Acacia greggii* —Photograph © 2006 by Kate Watters

Western Redbud, California Redbud *Cercis orbiculata*

PEA FAMILY (Fabaceae/Leguminosae) (*Cercis occidentalis* var. *orbiculata*)

Plants: Shrub to small deciduous tree, to 4 meters tall. **Leaves:** Alternate, nearly round, cordate at base, 3 to 10 centimeters wide. **Flowers:** In clusters, pink to magenta, pea-shaped. **Fruit:** Flattened pod. **Flowering Season:** March to June. **Elevation:** 1,200 to 7,500 feet.

Western redbud flowers in early spring before its glossy-green leaves appear. The drying, rust-colored pods dangle from the branches for many months as the seeds develop within. Natural germination of these impermeable seeds requires fire or flooding to scarify the hard seedcoat. In the Grand Canyon, redbud populations in the pre-dam high-water zone may be declining due to lack of scouring by spring floods.

Western redbud occurs in California, Nevada, Utah, and Arizona. In the Grand Canyon, this beauty graces shady, moist side canyons and slopes along the river. Although redbuds are frequently found along waterways, their drought tolerance allows them to also grow on dry hillsides and in rocky soils.

Native American people prune, cut to ground level, or burn redbud, which sprouts from the root crown, to encourage growth of straight, flexible branches for basketry. The early-blooming flowers provide nectar for a variety of pollinators. Solitary leafcutter bees (*Megachile* species) leave behind scalloped leaf margins after they tote off small pieces to use for nest construction. The pods can be roasted, rendering the seeds tasty. According to legend, Judas hung himself, after betraying Jesus, on the branch of a redbud tree and the flowers that previously were white, turned color or blushed with shame, hence another common name *Judas tree*.

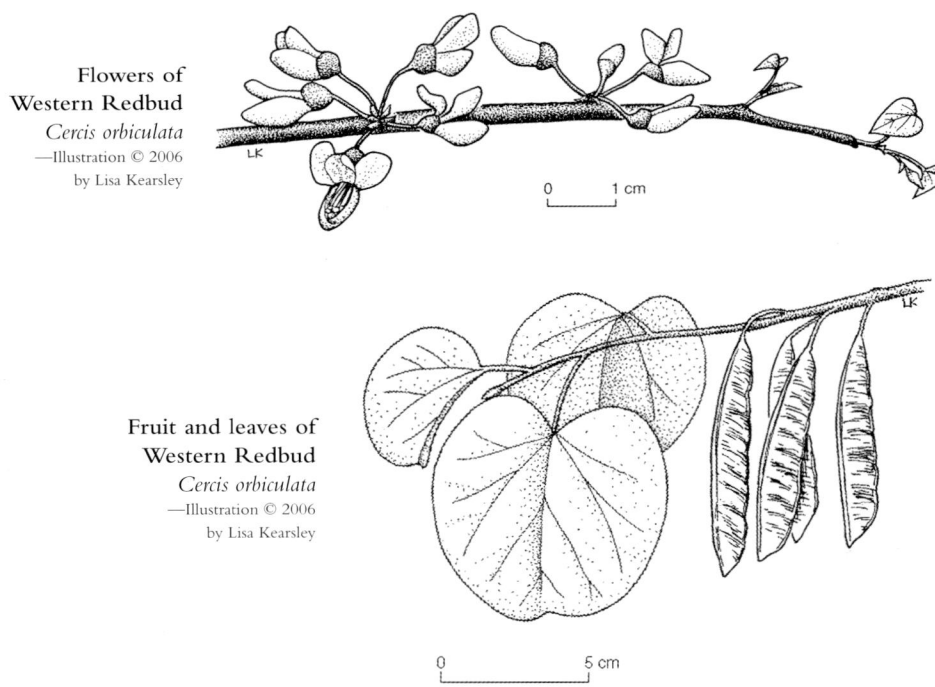

Flowers of
Western Redbud
Cercis orbiculata
—Illustration © 2006
by Lisa Kearsley

Fruit and leaves of
Western Redbud
Cercis orbiculata
—Illustration © 2006
by Lisa Kearsley

TREES • Pea Family

Western Redbud *Cercis orbiculata*
—Photograph © 2006 by Kate Thompson

Western Redbud *Cercis orbiculata*
—Photograph © 2006 by Kate Watters

Western Honey Mesquite

Prosopis glandulosa var. *torreyana*

PEA FAMILY (Fabaceae/Leguminosae)

(*P. juliflora*)

Plants: Broad-crowned deciduous tree, to 7 meters tall. **Leaves:** Bipinnate, with seven to seventeen leaflet pairs, to 17 centimeters long. **Flowers:** In cylindrical raceme that are up to 10 centimeters long; flowers greenish yellow, tiny. **Fruit:** Straw-colored pods. **Flowering Season:** April to August. **Elevation:** 1,200 to 3,900 feet.

Western honey mesquite is a common member of the community of plants that marks the pre-dam high-water line. Pre-dam seasonal peak flows on the Colorado River averaged over 100,000 cubic feet per second and prevented the establishment of permanent vegetation below the high-water line. Along the river, mesquite bosques form rounded hills as they stabilize sand dunes and talus slopes with their extensive root systems. A sturdy taproot delves deeper in search of groundwater.

Mesquite is often confused with catclaw acacia (*Acacia greggii*), which shares similar habitats. Mesquite bears straight, paired thorns, while acacia has sharp, curved prickles that resemble cat claws. The leaves of catclaw acacia are also much smaller and have fewer leaflets than those of mesquite.

Western honey mesquite occurs from California east to Texas and south to mainland Mexico and Baja California. In the Grand Canyon, it is found from River Mile 39 downstream to River Mile 76. In the Inner Gorge, the steep-walled region downstream of River Mile 76, catclaw acacia largely replaces mesquite, although scattered populations grow on tributary deltas and in side canyons. Mesquite is not common again in the river corridor until downstream of National Canyon, where it continues to Grand Wash Cliffs. John Wesley Powell may have been the first to record this species in the Grand Canyon, noting it in his journal that they encountered it near River Mile 39.

Tree cores from the Grand Canyon document ages of up to 750 years for western honey mesquite trees. Mesquite is thought to have coexisted with the mega-faunal grazers of the New World, such as mammoths and camels, that became extinct at the end of Pleistocene time. Today, deer, fox, and coyote eat the sweet, nutritious pods, and in doing so, carry the seeds away from the parent plant. Some seeds escape the digestive juices that scarify the

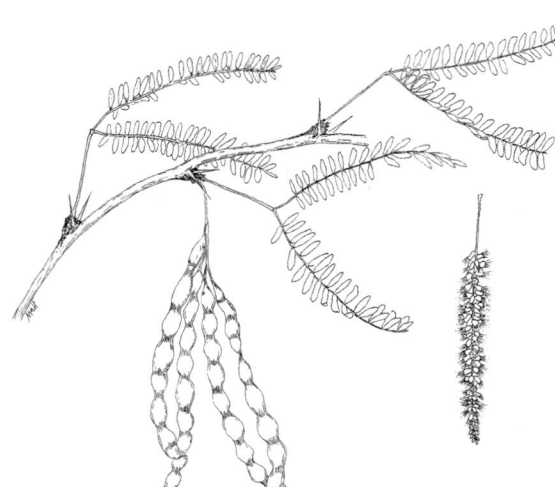

Fruiting branch and flowers of Western Honey Mesquite
Prosopis glandulosa var. *torreyana*
—Illustration © 2006 by Mar-Elise Hill

tough seed coat, and seeds are deposited in moist manure, an ideal microhabitat for germination. Seeds buried by rodents and ants also start new plants. Look for the tiny holes on the seedpods that are exit channels of parasitic Bruchid beetles (*Acanthoscelides* species) that eat the seeds during their larval stage. In the western Grand Canyon, the parasitic mistletoe (*Phoradendron californicum*) infests mesquite and catclaw acacia trees, robbing them of water and nutrients and often killing them.

For many cultures mesquite is key to their existence, providing fuel and food. The pods supply a dependable food source high in sugar and protein. Mesquite provides one of the richest sources of pollen and nectar for honeybees, resulting in delightful honey. The wood produces long-lived coals and imparts a savory aroma used to flavor barbecues. The demand for mesquite charcoal has led to the overharvest of the wood, but the pods provide the same aroma and could provide a more sustainable alternative.

Western Honey Mesquite *Prosopis glandulosa* var. *torreyana*
—Photograph © 2006 by Kristin Huisinga

Desert Mistletoe *Phoradendron californicum* —Photograph © 2006 by Celia Southwick

Tamarisk, Saltcedar *Tamarix ramosissima*
TAMARISK FAMILY (Tamaricaceae) (*T. chinensis*, *T. pentandra*)

Plants: Exotic, deciduous perennial shrub or tree, to 6 meters tall. **Leaves:** Tiny, scalelike. **Flowers:** In spikelike clusters, pink to white, with five minute petals. **Fruit:** Many-seeded capsule. **Flowering Season:** March to June. **Elevation:** 1,500 to 7,500 feet.

Tamarisk, an invasive, non-native tree, has quickly become a conspicuous element of the southwestern landscape and now occupies nearly all waterways. Its success is due in large part to the production of a thick layer of salty leaf litter below its dense canopy, inhibiting the survival of native species. In the spring and summer, minute pink to lavender blossoms drip from the branches like tassels. Although evergreen in appearance, the leaves fade to yellow ochre in autumn before sifting to the ground. Its penetrating root system allows it to usurp water from almost 10 meters below the surface. Tamarisk is able to use water that is not available to shallow-rooted plants, allowing it to survive in areas without regular, seasonal flooding. Although its tiny seeds are generally viable for only a few weeks, a typical tree produces about 600,000 seeds. You may notice a dense carpet of green, miniscule tamarisk seedlings on recently saturated drainage bottoms and river banks.

This fast-growing tree was originally introduced from Eurasia as an ornamental and to control erosion along waterways, field margins, and railroad tracks. It probably reached Utah in the early 1900s and became established in the Colorado River watershed by the 1930s. In the Grand Canyon, it is now a dominant tree species in the river corridor and in many side canyons. Beginning in 2002, a National Park Service control program, aided by hundreds of volunteers, has removed over 200,000 trees from the park's side canyons and tributaries. **Athel (*Tamarix aphylla*)**, a large, evergreen, invasive relative of tamarisk, is known only from a few locations in the Grand Canyon. Along roadsides and in irrigated fields in the Southwest, these trees can exceed 20 meters in height.

Fremont Cottonwood *Populus fremontii*
WILLOW FAMILY (Salicaceae)

Plants: Deciduous tree, to 30 meters tall. **Leaves:** Alternate, triangular, with toothed margins, 4 to 7 centimeters long. **Flowers:** In clusters; male and female catkins on separate trees; flowers inconspicuous. **Fruit:** Capsules. **Flowering Season:** March to May. **Elevation:** 1,200 to 5,200 feet.

On a clear blue summer day, you may hear the sound of a light rain, only to realize it is the leaves of Fremont cottonwood trembling in the wind. In spring, tiny clumps of cottonlike seeds float in the breeze. Many of the seeds may not become trees because they require consistently moist soil to germinate and the seedlings must withstand regular flooding. Statuesque cottonwood trees once defined riparian areas in the Southwest, and early explorers used them as indicators of surface water. Today, these trees are declining in areas where human activities are lowering water tables, and non-native species, such as tamarisk (*Tamarix* species), are encroaching into the habitat of these majestic giants.

Fremont cottonwood occurs throughout the Intermountain West from California east to Colorado and New Mexico. In the Grand Canyon, it grows in side canyons and at springs with permanent water, such as Stone, Tapeats, and Deer Creeks. These tributaries are good representations of what western riparian areas looked like prior to the 1900s. In the river corridor, you may occasionally see a Fremont cottonwood, but it may be gone the next time you pass by, perhaps cut down by a beaver.

Fremont cottonwood is planted in restoration areas because it is readily propagated from stem cuttings and is fast growing. The lightweight wood from the roots provides the material for Hopi kachina doll carvings. Native people use the twigs for basketry and consume immature flower clusters as chewing gum. Fremont cottonwood commemorates John Charles Fremont (1813–1890), an early western explorer, cartographer, and presidential candidate who collected plants along many of his mapping expeditions.

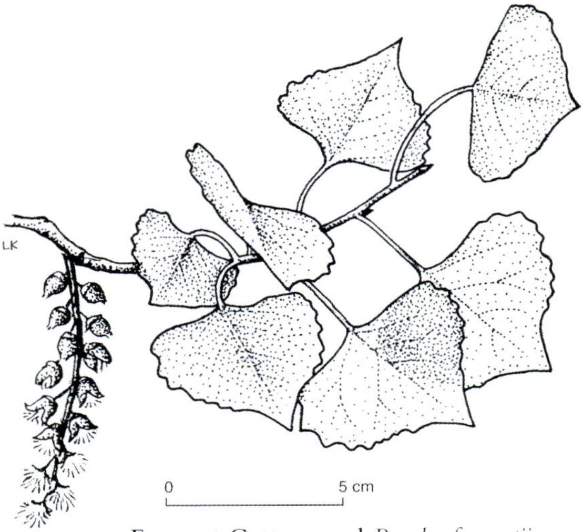

Fremont Cottonwood *Populus fremontii*
—Illustration © 2006 by Lisa Kearsley

Tamarisk *Tamarix ramosissima*
—Photograph © 2006 by Lynn Myers

Fremont Cottonwood *Populus fremontii* —Photograph © 2006 by David Edwards

Coyote Willow, Sand-Bar Willow *Salix exigua*
WILLOW FAMILY (Salicaceae)

Plants: Upright deciduous shrub to tree, to 7 meters tall. **Leaves:** Alternate, lanceolate, subtly-toothed, 2 to 11 centimeters long. **Flowers:** In axillary clusters that are 2 to 7 centimeters long; male and female catkins separate. **Fruit:** Capsule. **Flowering Season:** March to May. **Elevation:** 1,200 to 5,000 feet.

A pioneering species, coyote willow thrives on riverbanks that experience regular flooding. Flowing water transports broken stems or root fragments to new locations, where they colonize rapidly. The flowers appear in early spring just before or as the leaves begin to emerge. It would seem that their inconspicuous flowers would be wind-pollinated because they lack obvious visual clues to attract pollinators. Yet in fact, each flower in the catkin boasts nectaries that secrete a sweet odor, attracting many kinds of insects, especially bees. Pine-cone-shaped galls arise from the branch tips and house sawfly larva (*Euura* species). Coyote willow can be confused with the wide-leaved species of seep-willow (*Baccharis emoryi*, *B. salicifolia*), which coexist on moist shorelines. The seep-willows have wider, coarsely toothed leaves, obvious green striations on their woody stems, and flowers typical of the sunflower family.

Coyote willow boasts a great range, extending from Alaska to Louisiana and into Mexico. In the Grand Canyon, it is dominant along side creeks and in the river corridor, where it has become more abundant in the past thirty years, probably due to the lack of large floods.

Native people use the straight, young, flexible stems for basketry. Pre-Columbian inhabitants of the Grand Canyon crafted the pliable, red stems into split twig figurines. The willow bark and stems are rich in salicin, which is related to the original source of acetylsalicylic acid, the main ingredient in aspirin. The inner bark of coyote willow is a favorite food of beaver.

Goodding's Willow *Salix gooddingii*
WILLOW FAMILY (Salicaceae)

Plants: Deciduous tree, to 10 meters tall, with deeply fissured bark. **Leaves:** Alternate, lanceolate, bright green, shiny, 2 to 8 centimeters long. **Flowers:** In axillary clusters that are up to 7 centimeters long; male and female catkins separate. **Fruit:** Capsule. **Flowering Season:** April to May. **Elevation:** 1,200 to 7,700 feet.

A staple of healthy southwestern riparian habitats, Goodding's willow is one of the most regal trees in the Colorado River ecosystem, as well as Arizona's largest willow species. It is restricted to very wet areas. Its large canopy provides habitat for many bird species, including the endangered southwestern willow flycatcher (*Empidonax traillii extimus*). Goodding's willow is easy to distinguish from the more common coyote willow (*Salix exigua*) because it is a tall multiple-branched tree and has larger, bright green leaves.

Goodding's willow occurs from Kansas to California and south to Mexico. In the Grand Canyon, it grows in scattered populations along the length of the river, notably at Cardenas Marsh and at select locations downstream of Diamond Creek, where it forms dense stands on the fertile flood terraces. Perhaps the most famous individual, which some say resembles a dragon, resides at Granite Park and is culturally significant to the Hualapai people. A photograph from the 1923 U.S. Geological Survey expedition shows this tree at almost the same size as it is today, indicating that it is well over one hundred years old.

This willow species pays homage to botanist and conservationist, Leslie Newton Goodding (1880–1967), who explored and collected plants in the western United States. The genus name, *Salix*, is from Latin, meaning "to leap" or "to spring," in reference to its fast growth.

Coyote Willow
Salix exigua
—Photograph © 2006
by Kristin Huisinga

Goodding's Willow
Salix gooddingii
—Photograph © 2006
by Kate Watters

Goodding's Willow
Salix gooddingii
—Photograph © 2006
by Celia Southwick

SHRUBS AND FORBS

Century Plant *Agave utahensis*
AGAVE FAMILY (Agavaceae)

Plants: Succulent perennial, to 5 meters tall. **Leaves:** In a rosette, lanceolate, stiff, with marginal teeth, spine-tipped, to 50 centimeters long. **Flowers:** In panicles, yellow, to 3.5 centimeters long. **Fruit:** Three-parted capsule. **Flowering Season:** May to July. **Elevation:** 2,100 to 8,200 feet.

The century plant is one of the most unique plants to grace the precipitous slopes of the Grand Canyon. The common name is a misnomer because it only takes twenty to forty or more years for it to send up a towering, Dr. Seuss–like flowering stalk. Following this endeavor, the plant dies. Some century plants have underground stems, called rhizomes, that produce clones. You will sometimes see the next generation of smaller rosettes next to the larger parent rosettes. Stiff, succulent leaves spiral from the center of a rosette, fanning outward in perfect symmetry. The woody, burgundy-tinged spines arise from the leaf margins, and each leaf bears the shadowy imprint of the previous layer. These spines are distinctly different from the marginal fibers found on yucca species (*Yucca* species).

The Grand Canyon is home to three different century plants. Utah century plant (*Agave utahensis*) is the most common and is represented in the Grand Canyon by two subspecies: **Utah century plant (*A. utahensis* ssp. *utahensis*)** and **Kaibab century plant (*A. utahensis* ssp. *kaibabensis*)**. Kaibab century plant is the larger of the two and rarely produces clones from rhizomes, thus growing as a single plant. In the Grand Canyon, Kaibab century plant occurs from Marble Canyon downstream to Kanab Canyon, preferring calcareous or sandstone outcrops.

Utah century plant is a smaller, fewer-leaved version that produces numerous clones arranged in clumps. Its flowering stalks are up to 1 meter shorter than those of Kaibab century plant. It grows on open, rocky, usually limestone slopes in Arizona, California, Nevada, and Utah. In the Grand Canyon, plants are commonly found from the South Bass area downstream to Grand Wash Cliffs. While Utah century plant is found up to elevations of 4,600 feet, Kaibab century plant extends to 8,200 feet. Plants exhibiting characteristics of both species appear where the two subspecies overlap in distribution, suggesting that these subspecies hybridize.

A recently discovered species and an unfolding mystery to botanists, the **Grand Canyon century plant (*A. phillipsiana*),** is known only from a handful of sites within the Grand Canyon and is very distinct from the other agaves native to the Colorado Plateau. Its grayish green, arching leaves grow to nearly twice the length of those of Utah century plant. The Grand Canyon century plant produces a towering flower stalk dappled with creamy blossoms on long, widely spaced lateral branches. This species is believed to be an ancient, living cultivar—introduced and farmed by pre-Columbian people. These remarkable plants are found on terraces along a few major tributaries in the Grand Canyon, the majority of the plants in association with archaeological features such as roasting pits. The genetic clones that these agaves produce make them ideal for trade and cultivation. Grand Canyon century plant's closest relatives likely reside in northern Mexico. Pueblo inhabitants of the Grand Canyon probably acquired plants through trade with Mexico peoples.

A host of creatures including bees, hummingbirds, and bats pollinate agaves. Early southwestern inhabitants utilized century plants for food, fiber (for hairbrushes, sandals, blankets, and mats), medicine, and alcoholic and nonalcoholic beverages. Agave roasting, a traditional preparation still practiced in the Grand Canyon region by the Hualapai people, begins with the harvest of plants just prior to the emergence of the flower stalk, when the plant's core is engorged with sugars. The stem cores are roasted in a pit layered with coals, rocks, and moist vegetation, and covered with soil. Roasted agaves taste like dried papaya marinated or laced with molasses, making them sweet treats for ancient and modern desert dwellers. All agave species are protected by law, and you must obtain a permit for collection.

Grand Canyon Century Plant *Agave phillipsiana* —Photograph © 2006 by Kate Watters

Kaibab Century Plant
Agave utahensis ssp. *kaibabensis*
—Photograph © 2006 by David Edwards

Utah Century Plant
Agave utahensis ssp. *utahensis*
—Photograph © 2006 by Kyle George

Newberry's Yucca
Whipple Yucca (in part)
AGAVE FAMILY (Agavaceae)

Hesperoyucca newberryi
(*Yucca whipplei,* in part)

Plants: Succulent perennial, to 4 meters tall. **Leaves:** In a rosette, narrowly lanceolate, blue green, with minute teeth, sharp-pointed, to 1 meter long. **Flowers:** In terminal panicles, creamy white, bell-shaped, to 3 centimeters long. **Fruit:** Capsule. **Flowering Season:** February to June. **Elevation:** 1,500 to 4,000 feet.

The sword-shaped, blue green leaves and massive creamy white flowering stalks of Newberry's yucca are conspicuous and distinctive in the desert landscape. Each plant requires six to seven years to accumulate enough food reserves to produce its inflorescence. The rosette dies after the plant flowers, a trait not shared by yuccas (*Yucca* species). Botanists are currently deliberating as to whether Newberry's yucca is more closely related to yuccas or false yuccas (*Hesperaloe* species).

Plants in the genus *Hesperoyucca* primarily occur in chaparral, coastal, and desertscrub communities of California with fewer locations in the deserts of Arizona and Baja California. In the Grand Canyon, look for Newberry's yucca on shelves and steep, rocky slopes, from Stone Creek downstream sporadically to Grand Wash Cliffs. Fossil evidence suggests that this species was once more continuously distributed during Pleistocene time, but it is absent from pack rat middens in the Grand Canyon from the same time period. Botanists do not know when and how this species arrived in this region.

One species of yucca moth (*Tegeticula maculata*) is the only pollinator of Newberry's yucca, and the developing seeds provide the larvae's only food source. Despite the large inflorescence with numerous flowers, only about 10 percent of the plant's flowers escape predation and produce mature fruit. Where agave (*Agave* species) is absent in the landscape, it is believed that indigenous groups roasted the hearts of *Hesperoyucca* plants, which provided a food source similar to agave. Newberry's yucca is named for John Strong Newberry (1822–1892), an American botanist and paleontologist, who first collected this plant at the confluence of Diamond Creek and the Colorado River. He accompanied Lieutenant Joseph Ives on an exploratory, upriver expedition in a steamboat.

YUCCA
AGAVE FAMILY (Agavaceae)

Yucca species

There are about forty species of yucca known worldwide. Many are economically and culturally important, especially for Native Americans in the western United States who use the fiber for cordage, matting, fishing nets, and clothing. The fruit and roots also provide food and ceremonial items. Because yuccas spread vegetatively through aboveground stem growth, Native Americans are able to sustainably harvest whole rosettes.

When not in bloom, yucca can be confused with century plant (*Agave* species). However, yucca has long fibers that curl away from the leaf edges while century plant bears barbed spines on its leaf margins. **Joshua-tree (*Yucca brevifolia*)** is an exception to the rule, with very small, sharp teeth on the leaf margins instead of fibers. Joshua-tree, a magnificent plant with a treelike habit, is believed to be an ancient form from which other species within the genus *Yucca* evolved. Today, plants are restricted to the Mojave Desert. They grow only in the extreme southwestern part of Grand Canyon National Park, in limestone soils at the top of Grand Wash Cliffs.

Newberry's Yucca *Hesperoyucca newberryi*
—Photograph © 2006 by Glenn Rink

Joshua-Tree *Yucca brevifolia* —Photograph © 2006 by Kristin Huisinga

Yuccas are self-incompatible, requiring pollen from other plants for successful fruit set, which has lead to a mutualistic relationship with yucca moths (*Tegeticula* species). The female moth packs the openings of the stigma with a mass of sticky pollen gathered from other flowers. She then pierces the ovary with her ovipositor and lays her eggs inside. Moth larvae and seeds develop together, with the larvae feeding on the immature seeds. When fully developed, the larvae chew their way through the ovary wall and escape. Some seeds fully mature, but herbivores and lack of rain take their toll on seedlings.

Banana Yucca *Yucca baccata*
AGAVE FAMILY (Agavaceae)

Plants: Succulent perennial, to 1.3 meters tall. **Leaves:** In a rosette, lanceolate, pale green to bluish green, with marginal fibers, sharp-pointed, to 90 centimeters long. **Flowers:** In panicles, white to cream, to 15 centimeters long. **Fruit:** Fleshy, banana-shaped capsule. **Flowering Season:** April to May. **Elevation:** 1,700 to 7,200 feet.

After a winter with ample moisture, banana yucca sends up stout stalks heavy with waxy, bell-shaped blossoms. Unlike many yuccas, the stalks rarely surpass the long, wide leaves. Banana yucca occurs from California east to Colorado and Texas. In the Grand Canyon, plants are uncommon in the river corridor and primarily inhabit dry, rocky slopes and upper benches of side canyons, extending through blackbrush and piñon-juniper communities to the forested rims.

Banana yucca's fleshy, edible fruits may have evolved for dispersal by large mammals that lived during Pleistocene time, most of which are now extinct. Rodents and rabbits are now the primary seed dispersers. Native American people use banana yucca extensively for food, shampoo, and fiber. The leaf fibers are used to make sandals, baskets, cordage, and fabric. Ethnobotanists suggest that the exceptionally long-leaved banana yucca population at Stone Creek may have been nurtured by indigenous people because of the plant's longer, superior fiber.

Soaptree Yucca *Yucca elata*
AGAVE FAMILY (Agavaceae)

Plants: Succulent perennial, with well-developed trunk, to 9 meters tall. **Leaves:** In a rosette, linear, rigid, with fine marginal fibers, sharp-pointed, to 90 centimeters long. **Flowers:** In long-stalked panicles, white to cream, 3 to 5 centimeters long. **Fruit:** Dry, three-parted capsule. **Flowering Season:** May to June. **Elevation:** 1,800 to 7,400 feet.

Even when soaptree yucca lacks its flowering stalk, it has a distinctive rosette of sword-like leaves atop a shaggy, well-developed trunk. Most other narrow-leaved yuccas lack trunks and the flowers are primarily arranged in short-stalked racemes. Some yucca plants in the canyon may be a hybrid between soaptree yucca and **narrow-leaved yucca (*Yucca angustissima*)**, exhibiting characteristics of both. Plant taxonomists still struggle with the peculiarities of structure manifested by yuccas in the Grand Canyon.

Soaptree yucca is found from southwestern Texas and west to Arizona. In the Grand Canyon, it is especially noticeable on desert slopes and dry soils along the river from Badger Rapids at River Mile 8 to Havasu Creek.

The common name, *soaptree yucca*, describes the saponin-rich stems and roots that lather when mixed with water and are used as soap and shampoo by native people.

Soaptree Yucca hybrid with raceme *Yucca elata* —Photograph © 2006 by John Dewine

Typical form of Soaptree Yucca with long-stalked panicle *Yucca elata* —Photograph © 2006 by Joe Pollock

Banana Yucca *Yucca baccata* —Photograph © 2006 by Bronze Black

Tumble Pigweed, Pale Amaranth   *Amaranthus albus*
AMARANTH FAMILY (Amaranthaceae)

Plants: Exotic, many-branched, annual herb, to 1 meter tall and often as wide. **Leaves:** Alternate, spatulate to obovate, green, reddish purple below, 1 to 7 centimeters long, veins prominent. **Flowers:** In axillary clusters, inconspicuous flowers, with shiny bracts. **Fruit:** Tiny black seeds. **Flowering Season:** May to November. **Elevation:** 1,800 to 7,000 feet.

Tumble pigweed, a robust annual, emerges in disturbed areas following summer rains. As the flowers mature, they develop into thousands of shiny, black seeds encased in spiny bracts. The common name *tumble pigweed* refers to the tumbling behavior of mature plants blowing in the wind. Seeds disperse as the plants tumble.

Tumble pigweed is native to Latin America. The first documentation within Grand Canyon National Park was near Kanab Creek in 1994, but this species is widespread in other locations within the Inner Canyon and on the rims.

Plants in the genus *Amaranthus* have been cultivated as crop plants worldwide. The seeds and greens are a good source of protein, iron, calcium, and Vitamins A and C. Ground seeds are made into porridge and cakes. For thousands of years, native people in the Four Corners area have been harvesting both wild and cultivated amaranth.

Hopi people cultivated a **red amaranth (*A. cruentus*)** for a brilliant pink food dye, and some plants still persist in Hopi fields despite its replacement by commercial dyes. In Havasu Canyon, a long-inhabited location, six other amaranth species thrive in old Havasupai fields and garden areas. These patterns exemplify the influence of native people on distribution, survival, and diversity of wild plants. Were these species actively sought out from other regions for cultivation or passive arrivals that were merely tolerated on disturbed field margins?

Red Barberry   *Mahonia haematocarpa*
BARBERRY FAMILY (Berberidaceae)   (*Berberis haematocarpa*)

Plants: Evergreen, perennial shrub, to 4 meters tall. **Leaves:** Alternate, pinnately compound with three to seven leaflets, terminal leaflet longer, bluish green, spiny-toothed, leathery. **Flowers:** In loose clusters, yellow, to 8 millimeters wide. **Fruit:** Small, red berry. **Flowering Season:** February to June. **Elevation:** 2,000 to 4,500 feet.

Even without its bright red berries, the bluish green hue of red barberry's foliage offers an attractive contrast against the desert landscape. The evergreen, spine-tipped leaves are sure to keep all passersby at a respectable distance. In the early spring when most other shrubs are still dormant, its clusters of bright yellow flowers are a welcome surprise.

Red barberry occurs from Texas and Colorado west to California. In the Grand Canyon, red barberry grows on slopes, along trails, and in side canyons, but not along the main river corridor. Look for it in Saddle and Matkatamiba Canyons and along the Bright Angel Trail.

Many people mistake this shrub for its close relative **Fremont barberry (*Mahonia fremontii*)**, which grows at higher elevations, has bluish black berries, and more-or-less equal-sized terminal and lateral leaflets. Similar in stature, shrub live oak (*Quercus turbinella*) bears stiff, hollylike leaves and also grows in the Inner Canyon. The leaves of shrub live oak are simple and olive green, compared to the compound, bluish green leaves of red barberry.

Birds and small mammals relish the edible berries. Although the leaves can be poisonous, deer may browse red barberry when food is scarce. The stems and roots produce a brilliant, yellow dye used to color cloth, buckskins, and baskets. All barberries contain berberine, a compound that is therapeutic for the liver and used to be an ingredient in eye drops.

A small Tumble Pigweed *Amaranthus albus*
—Illustration © 2006 by Mar-Elise Hill

Leafy branches of a robust Tumble Pigweed
Amaranthus albus —Photograph © 2006 by Max Licher

Red Barberry *Mahonia haematocarpa*
—Photograph © 2006 by Roger Dale

Scarlet Lobelia, Cardinal Flower *Lobelia cardinalis*
BELLFLOWER FAMILY (Campanulaceae)

Plants: Perennial herb, to 1 meter tall. **Leaves:** Alternate, linear to lanceolate or ovate, dark green, toothed, to 15 centimeters long. **Flowers:** In spikelike racemes, red, to 1.5 centimeters long. **Fruit:** Capsule. **Flowering Season:** June to October. **Elevation:** 1,800 to 4,500 feet.

In the late summer, you might encounter the sudden flash of scarlet lobelia blossoms amidst crowds of riparian foliage in many side canyons. It has a distinctive, asymmetrical flower shape characteristic of some members of the bellflower family, with two corolla lobes above and three below. This shape distinguishes it from the blooms of cardinal monkeyflower (*Mimulus cardinalis*), which also has reddish flowers and shares scarlet lobelia's affinity for flowing water.

Scarlet lobelia is widely distributed throughout North and Central America. In the Grand Canyon, it grows along streams and in seeps and springs the length of the river.

The showy flowers attract hummingbirds that drink the nectar and spread pollen from bloom to bloom. Some of the species in this genus contain alkaloids with properties similar to those of nicotine. Leaf extracts and fruits are poisonous, leading many modern herbalists to discourage its use.

The genus name honors Flemish botanist Matthias de L'Obel (1538–1616). The specific epithet, *cardinalis*, describes the brilliant scarlet color of the flowers, similar to that of cardinals' robes.

Glandular Threadplant *Nemacladus glanduliferus* ssp. *arizonica*
BELLFLOWER FAMILY (Campanulaceae)

Plants: Winter annual herb, 5 to 25 centimeters tall. **Leaves:** Oblanceolate to elliptical, to 15 millimeters long. **Flowers:** In racemes, with a zigzagged, threadlike axis; flowers white, to 2.5 millimeters long. **Fruit:** Capsule. **Flowering Season:** March to May. **Elevation:** 1,200 to 4,000 feet.

Barely noticeable with their threadlike stems, glandular threadplant boasts gorgeous but tiny flowers. Seeds germinate during years with good winter rain. You have to bend down low to admire the maroon-tipped, white flowers of this tiny plant. Many plants in this genus have strange crystalline rods on their filaments, the threadlike stalk of the flower stamen. Botanists are not sure what the rods are but believe they may be serving as an enticement for pollinators.

Glandular threadplant is endemic to the southwestern United States and occurs from New Mexico and Utah, west to California. In the Grand Canyon, plants grow on slopes and flats along the river and in side canyons the length of the river.

Scarlet Lobelia *Lobelia cardinalis* —Photograph © 2006 by David Edwards

Glandular Threadplant *Nemacladus glanduliferus* ssp. *arizonica*
—Photograph © 2006 by Kristin Huisinga

Sandpaper-Bush, Utah Mortonia
BITTERSWEET FAMILY (Celastraceae)

Mortonia utahensis
(*Mortonia scabrella* var. *utahensis*)

Plants: Erect shrub, to 2 meters tall. **Leaves:** Alternate, ovate to round, yellow green, with thick margins, leathery, 6 to 16 millimeters long. **Flowers:** In terminal panicles, white, inconspicuous. **Fruit:** Capsule. **Flowering Season:** March to September. **Elevation:** 1,400 to 4,800 feet.

Sandpaper-bush often grows in pure stands. The nearly vertical arrangement of its small, rough leaves with thickened margins helps to deflect the sun's rays, minimizing water loss. Like its tree-size relative, crucifixion-thorn (*Canotia holacantha*), also in the bittersweet family, its stems photosynthesize, another means of adapting to a xeric environment.

Sandpaper-bush occurs in and around the fringes of the Mojave Desert in Arizona, Utah, Nevada, and California. In the Grand Canyon, it prefers limestone or sandstone microhabitats and rocky slopes from Havasu Canyon downstream to Grand Wash Cliffs.

A smaller-leaved species, **Rio Grande saddlebush (*Mortonia scabrella*)**, is geographically isolated from sandpaper-bush, restricted to the very southern reaches of Arizona, and into western Texas and northern Mexico. Pack rat middens suggest that both species were once more abundant. Our species was possibly even a dominant component of the landscape between 10,000 and 8,200 years ago in the western part of the Grand Canyon and southern Nevada.

A variety of insects, including chrysomelid beetles, wasps, bees, and butterflies visit the tiny flowers. The genus name honors Samuel George Morton (1799–1851), an American physician and naturalist.

Common Fiddleneck
BORAGE FAMILY (Boraginaceae)

Amsinckia menziesii var. *intermedia*

Plants: Annual herb, 20 to 100 centimeters tall. **Leaves:** Alternate, linear to narrowly lanceolate, hairy, 2 to 15 centimeters long. **Flowers:** In terminal spikes; yellow orange fused petals, with red orange marks; flowers to 1 centimeter long. **Fruit:** Triangular nutlet. **Flowering Season:** March to June. **Elevation:** 1,200 to 7,000 feet.

After wet winters, common fiddleneck forms dense stands of finely hairy plants that feel like fiberglass to the touch. The tiny, yellowish orange spirals of common fiddleneck flowers are self-pollinated and can produce viable seed with or without the service of pollinators. Flowers unfurl so that the freshest blooms are prominently positioned to allow easy visitation. The seeds and leaves contain alkaloids, which are toxic to other plants and herbivores, giving common fiddleneck a competitive advantage. This plant is also called *devil's lettuce* because of the coating of spiny hairs.

The flower structure of common fiddleneck bears a striking resemblance to the purple-flowered scorpionweed (*Phacelia* species). However, common fiddleneck has yellow flowers and a unique fruit shape, a nutlet.

Common fiddleneck is widespread from British Columbia, throughout the western United States, and into mainland Mexico and Baja California. In the Grand Canyon, it grows in side canyons and upper beaches along the river from Nankoweap Canyon downstream to Grand Wash Cliffs.

Sandpaper-Bush *Mortonia utahensis*
—Photograph © 2006 by Glenn Rink

Sandpaper-Bush *Mortonia utahensis*
—Photograph © 2006 by Kate Watters

Common Fiddleneck *Amsinckia menziesii*
var. *intermedia* —Photograph © 2006 by Kristin Huisinga

Common Fiddleneck *Amsinckia menziesii*
var. *intermedia* —Photograph © 2006 by Glenn Rink

Cryptantha
BORAGE FAMILY (Boraginaceae)

Cryptantha species

Plants: Erect annual or perennial herbs, 10 to 80 centimeters tall. **Leaves:** Linear to oblanceolate, bristly hairy, to 4 centimeters long. **Flowers:** In densely packed, bractless spikes or racemes; fused, white petals. **Fruit:** Four nutlets, variously grooved or winged. **Flowering Season:** February to September. **Elevation:** 1,200 to 7,000 feet.

The cryptanthas belong to one of the largest genera of plants in the Grand Canyon, comprising over two dozen species. You can recognize members of the genus from a combination of the bristly hairs that cover the stems and leaves and the coiled arrangement of the four-petaled, white flowers. The most important diagnostic feature is the arrangement of four fruits, or nutlets, in each flower, which are often equipped with barbs or hooks.

Even for botanists, individual species of cryptanthas are some of the more challenging plants to distinguish because they require microscopic inspection of the fruit. Some species are large, coarse, short-lived perennials, such as **woody cryptantha (*Cryptantha racemosa*)**, which usually extends to higher elevations. In arid, lower elevation areas, most cryptanthas are delicate annuals that range from a few centimeters to half a meter tall. Two other look-alike plants also have four-petaled flowers arranged in a coiled pattern and hairy stems and leaves: common fiddleneck (*Amsinckia menziesii* var. *intermedia*) has yellow flowers and bristly hairs, and scorpionweed (*Phacelia* species) has purple flowers and glandular hairs.

Cryptanthas are widespread throughout the desert Southwest and extend from Washington south to Texas and Mexico. In the Grand Canyon, they grow along sandy washes and in gravelly slopes the length of the river.

The bristly fruits cling to passing animals, as well as clothing, thereby aiding seed dispersal. Because cryptantha flowers are so small, very few insects are capable of collecting their nectar and pollen. As a result, most cryptanthas are self-pollinating and some are even *cleistogamous*, which means the flowers never open. The genus name, *Cryptantha*, is from the Greek words *cryptos* and *anthos*, meaning "hidden flower," which describes the tiny, self-fertilizing flowers.

Matted Tiquilia
BORAGE FAMILY (Boraginaceae)

Tiquilia latior

(*Coldenia hispidissima* var. *latior*)

Plants: Mat-forming perennial herb, 20 to 60 centimeters wide. **Leaves:** Alternate, linear to narrowly lanceolate, with rigid hairs, to 1 centimeter long. **Flowers:** Solitary in leaf axils, white to pink, 5 to 8 millimeters long. **Fruit:** Nutlet. **Flowering Season:** April to September. **Elevation:** 1,200 to 6,000 feet.

Matted tiquilia forms characteristically low, spreading mats of bristly hairy leaves, arising from a woody taproot. Look close among the densely matted leaves to find its whitish pink, funnel-shaped flowers, typical of the borage family.

Matted tiquilia is endemic to the Colorado Plateau, growing in cool desertscrub and piñon-juniper communities. In the Grand Canyon, it is widespread on rocky beaches, dunes, gravel bars, and dry slopes throughout the Inner Canyon.

Equally common in the Grand Canyon, **shrubby coldenia (*Tiquilia canescens*)**, has wider, more oval leaves that are covered in soft, white hairs. This species is found on dry, sunny mesas and slopes, preferring the rocky, calcareous soils of the Inner Canyon.

The genus name, *Tiquilia*, is a South American indigenous name for flowers in this genus. The specific epithet, *latior*, means "low" or "prostrate."

An annual species of Cryptantha *Cryptantha* species
—Photograph © 2006 by Kristin Huisinga

Inset —Photograph © 2006 by David Inouye

Matted Tiquilia *Tiquilia latior*
—Photograph © 2006 by Max Licher

Matted Tiquilia *Tiquilia latior*
—Photograph © 2006 by Max Licher

Shrubby Coldenia *Tiquilia canescens*
—Photograph © 2006 by Max Licher

Desert Broomrape *Orobanche cooperi*
BROOMRAPE FAMILY (Orobanchaceae)

Plants: Annual root parasite, 10 to 40 centimeters tall. **Leaves:** Alternate, reduced and scalelike. **Flowers:** In spikelike clusters, dull purple to yellow. **Fruit:** Capsule. **Flowering Season:** February to September. **Elevation:** 2,100 to 7,000 feet.

The fleshy, unbranched stem of desert broomrape cracks the soil when it emerges, displaying an abundance of unforgettable purple flowers. There is not a speck of green on this plant, and you will often find it dusted with a layer of sand. Lacking chlorophyll and well-developed leaves, these plants rely on other species to survive. Desert broomrape robs nutrients from the roots of fetid marigold (*Thymophylla pentachaeta* var. *pentachaeta*), coyote willow (*Salix exigua*), brittlebush (*Encelia* species), and white bursage (*Ambrosia dumosa*), among others. In fact, if you see a tiny seedling alone on an open sandy area, it will likely shrivel and die because it lacks a host plant.

Desert broomrape occurs from California east to Texas. In the Grand Canyon, it is found on beaches in the river corridor and in forested areas on the South Rim.

Widely distributed, broomrapes have many medicinal uses. Broomrapes can damage crops with their sneaky nutrient-robbing lifestyle.

Graythorn *Ziziphus obtusifolia* var. *canescens*
BUCKTHORN FAMILY (Rhamnaceae)

Plants: Intricately branched shrub, to 4 meters tall. **Leaves:** Alternate, oblong to ovate, gray green, to 5 millimeters long. **Flowers:** In flat-topped clusters, whitish green, tiny. **Fruit:** Dark blue drupe. **Flowering Season:** April to September. **Elevation:** 1,200 to 6,400 feet.

Graythorn is easy to pick out of the well-armed desert crowd with its ashy, grayish green branchlets arranged at right angles and terminating in sharp, rigid points. Graythorn blooms throughout the growing season, whenever rainfall is adequate, and sheds its reduced leaves to conserve resources during dry spells.

Try not to confuse this with another thorny-branched shrubby tree, crucifixion-thorn (*Canotia holacantha*), which is nearly leafless and has dry capsules. Graythorn has well-developed leaves and dark blue drupes.

Graythorn occurs from southeastern California east to Texas and south to Mexico. In the Grand Canyon, it grows on dry slopes and flats from Saddle Canyon downstream to Grand Wash Cliffs. Look for graythorn at Deer Creek Valley and Havasu Canyon.

Foxes, raccoons, ringtails, and a variety of birds savor the fruit, which is not very tasty to humans. Graythorn thickets provide important protection for birds that build their nests in the dense branches. Native American tribes use a root decoction to treat eye pain and to make a medicinal soap.

Desert Broomrape *Orobanche cooperi*
—Photograph by Lori Makarick/National Park Service

Graythorn *Ziziphus obtusifolia* var. *canescens*
—Photograph © 2006 by Max Licher

Graythorn *Ziziphus obtusifolia* var. *canescens* —Photograph © 2006 by Kate Watters

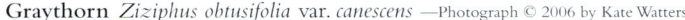

BUCKWHEAT *Eriogonum* species
BUCKWHEAT FAMILY (Polygonaceae)

There are over twenty-five buckwheat species in the Grand Canyon, and while you may not learn them all, you will recognize many as members of the buckwheat family by their tiny flowers gathered into small, pom-pom clusters. Desert trumpet (*Eriogonum inflatum*)is probably the most common species in the Grand Canyon with **skeleton weed (*E. deflexum*)** and **shrubby wild buckwheat (*E. corymbosum*)** also conspicuous. The annual skeleton weed has a basal rosette and structure that resembles desert trumpet, but its white flowers dangle like bells from its flat-topped, airy flowering stalks. It prefers habitats similar to those of desert trumpet. Shrubby wild buckwheat is a large, rounded shrub, with white or yellow flowers covering the entire plant in the fall. Look for it at Lees Ferry. **Heermann's wild buckwheat (*E. heermannii* var. *subracemosum*)** is a more rare, densely branched, pale gray shrub that grows on limestone outcrops.

Members of the buckwheat family are major components of the southwestern desert flora and are essential for butterfly survival. Native people made a tea to cure colds and fashioned the hollow stems into tobacco pipes. The Greek word *Erion* means "wool" and *gonu* means "knee," referring to the woolly leaves and swollen joints of many of the plants in this genus.

Shrubby Wild Buckwheat
Eriogonum corymbosum
—Photograph © 2006
by Kristin Huisinga

Heermann's Wild Buckwheat
Eriogonum heermannii var. *subracemosum*
—Photograph © 2006
by David Edwards

Skeleton Weed
Eriogonum deflexum.
—Photograph © 2006 by Wendy Hodgson

Desert Trumpet *Eriogonum inflatum*

Plants: Annual or perennial herb, to 1 meter tall. **Leaves:** In basal rosette, ovate to round, dark green with wavy margins, coarsely hairy, 1 to 5 centimeters long. **Flowers:** In flat-topped, diffuse panicles, reddish yellow, 1 to 3 millimeters long. **Fruit:** Three-angled achene. **Flowering Season:** March to October. **Elevation:** 1,200 to 4,200 feet.

Desert trumpet is one of the more conspicuous species in the buckwheat family because many, if not all, of its smooth, grayish green stems are inflated just below the nodes. Moth larvae that reside inside the hollow stems cause the swelling, although some stems may inflate without larvae. Desert trumpet utilizes winter moisture to produce a rosette of leaves, usually reserving flowering for the summer monsoon period. The plants dry out and often remain as skeletons for a year or more, making them easy to recognize when not in flower.

Desert trumpet grows in warm desertscrub and piñon-juniper communities from California east to Colorado and south to Mexico. In the Grand Canyon, it lives along the entire river corridor, in side canyons on rocky slopes, and in washes.

Desert Trumpet
Eriogonum inflatum
—Photograph © 2006 by Kristin Huisinga

Inset: Desert Trumpet
Eriogonum inflatum
—Photograph © 2006 by David Inouye

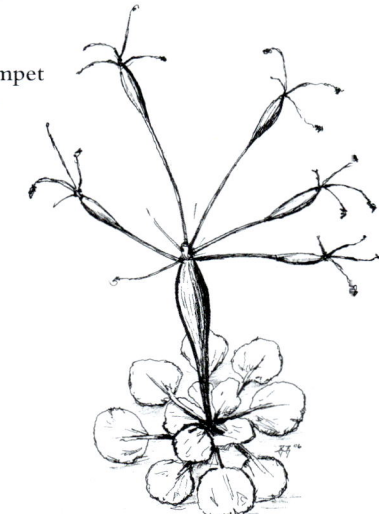

Desert Trumpet *Eriogonum inflatum*
—Illustration © 2006 by Bronze Black

Desert Windflower
Anemone tuberosa
BUTTERCUP FAMILY (Ranunculaceae)

Plants: Many-stemmed, perennial herb, 8 to 50 centimeters tall. **Leaves:** Mostly basal, with three deeply lobed leaflets, 3 to 16 centimeters long. **Flowers:** One or two on long stalks, whitish pink to purple petal-like sepals that are up to 1.4 centimeters long. **Fruit:** Densely woolly achene. **Flowering Season:** February to May. **Elevation:** 1,200 to 5,700 feet.

The precisely divided leaves and graceful pink flowers of desert windflower make it easy to identify. Because its leaves are not adapted to the harsh desert conditions, it takes advantage of the cool, spring temperatures and moist soil by blooming early in the year. Deep beneath the surface it hides a tuberous root rich in carbohydrates that feeds the growing plant. If you miss its beautiful elusive flower, look for the headlike clusters of woolly fruits above the conspicuous leaves.

Desert windflower is found in southern California, Nevada, and Utah, and east to Texas. In the Grand Canyon, it grows in shady microhabitats on desert slopes and shelves or among rocks in dry washes from around South Canyon downstream to Grand Wash Cliffs. It also inhabits volcanic cinder soils on the western North Rim.

Other species in this genus are medicinal, offering relief to weary souls who suffer from insomnia and nervousness. In Greek mythology, the genus name, *Anemone*, is from Anemos, the Wind, who sends the delicate anemone flowers as a signal of the coming of spring winds. An unknown poet wrote:

> *Coy anemone that ne'er uncloses*
> *Her lips until they're blown on by the wind.*

Golden Columbine
Aquilegia chrysantha
BUTTERCUP FAMILY (Ranunculaceae)

Plants: Perennial herb, 30 to 120 centimeters tall. **Leaves:** Divided into three parts, deeply lobed, bluish green, to 45 centimeters long. **Flowers:** One to many, canary yellow, with spreading petal-like sepals; flowers to 10.5 centimeters long. **Fruit:** Many-seeded follicle. **Flowering Season:** April to September. **Elevation:** 1,200 to 8,000 feet.

Golden columbine buries its nectar deep in the tip of the flower's backward-projecting spurs, ensuring that only the long proboscises of butterflies and the long tongues of hummingbirds can reach it. In their effort to acquire the nectar, each pollinator, as it brushes up against the anthers, maximizes the transfer of pollen. In turn, golden columbine rewards pollinators enough energy and incentive to travel onward to the next nectar-laden flower. Bees, which have short proboscises, bite through the spur to rob nectar, avoiding the duty of pollination. To prevent self-pollination, the stamens mature first, shedding their pollen before the style emerges at the center of the flower.

Golden columbine occurs from Colorado south to Texas, Arizona, and Mexico. In the Grand Canyon, it grows in side canyons and seeps throughout the river corridor and on the North Rim. Boasting one of the greatest elevational ranges of any Arizona plant species, it occurs from river level at 1,200 feet to the top of the San Francisco Peaks (near Flagstaff) at 12,000 feet.

Golden columbine is one of seven native columbines that grow in Arizona, three of which are found in the Inner Canyon. **Alcove columbine (*Aquilegia micrantha*)** is usually shorter than golden columbine and in the Grand Canyon is restricted to a few side canyons.

It has very small white, cream, blue, or pink flowers and sticky leaves. **Desert columbine (*A. desertorum*)**, common at Buck Farm Canyon, is also shorter in stature and has pinkish red petals tipped with yellow. Hybrids between these two species are also likely given the variety of orange, yellow, and pink blossoms on Grand Canyon plants.

The genus name, *Aquilegia*, may be derived from the Latin *aquila*, meaning "eagle," referring to the resemblance of the curved spur to an eagle's talon. Or it may originate from the Greek *aqua* and *legere*, meaning "to draw water," possibly describing its affinity for wet habitats.

Desert Windflower *Anemone tuberosa*
—Photograph © 2006 by Max Licher

Golden Columbine *Aquilegia chrysantha*
—Photograph © 2006 by Kathy Darrow

Columbine (*Aquilegia* species) displaying two flower colors —Photograph © 2006 by Lisa A. Hahn

Desert Larkspur *Delphinium parishii*
BUTTERCUP FAMILY (Ranunculaceae)

Plants: Erect perennial herb, to 1.2 meters tall. **Leaves:** Alternate, deeply palmately lobed, to 7.5 centimeters long. **Flowers:** In racemes, blue to deep purple, spurred, to 2.5 centimeters long. **Fruit:** Capsule. **Flowering Season:** May to June. **Elevation:** 1,200 to 3,800 feet.

Desert larkspur's stalks of purple or blue flowers stand tall above other greenery in springtime. The exquisite spurred blossoms conceal nectar deep within, where only pollinators with long proboscises can obtain it. As spring turns to summer, delicate capsules with three distinct sections dry and turn tan.

Desert larkspur occurs in California, Nevada, Utah, and Arizona. In the Grand Canyon, plants grow on steep slopes and open canyons from Pipe Creek to Grand Wash Cliffs. Another higher-elevation Grand Canyon species, **barestem larkspur (*Delphinium scaposum*)**, lacks leaves along its stem and bears only a basal rosette of leaves.

All *Delphinium* species are extremely toxic and should not be consumed. The genus name, *Delphinium*, is derived from the Greek word *delphinos* for "dolphin," descriptive of the arching nectar spurs.

Teddy-Bear Cholla *Cylindropuntia bigelovii* var. *bigelovii*
CACTUS FAMILY (Cactaceae) (*Opuntia bigelovii*)

Plants: Cylindrical, jointed, succulent shrub, to 2 meters tall. **Spines:** Interlaced, three to eleven per areole, to 3.5 centimeters long. **Flowers:** Whitish to pale green, sometimes tipped red, to 2.5 centimeters long. **Fruit:** Yellow to green dry berry. **Flowering Season:** March to June. **Elevation:** 1,200 to 2,900 feet.

Teddy-bear cholla may look fuzzy and cuddly, but it is far from it! Its dense, interlacing spines present a mixed blessing. They limit photosynthesis by obscuring the stem, but also protect the plants from animals that may want to eat the juicy contents. During the summer, the spines provide shade and radiate heat away from the plant, while in winter they buffer the plant from freezing temperatures. Lacking fertile seeds, teddy-bear cholla reproduces vegetatively. Its spiny joints readily detach, cling to passing animals, and once fallen, root to create new populations. If one attaches to you, use a comb to lift it off.

Teddy-bear cholla's range closely follows the more arid parts of the Sonoran Desert as far south as Baja California and Sonora, Mexico. It is also a common species in the Mojave Desert, with its northern populations reaching extreme southern Nevada and the Grand Canyon. In the Grand Canyon, it grows on gravelly flats and slopes from 192 Mile Canyon to Grand Wash Cliffs. It is especially abundant in the Diamond Creek area.

Another species, which may be a hybrid form of teddy-bear cholla, is **Peach Springs Canyon cholla (*Cylindropuntia abyssi*)**. It is known only from Peach Springs Canyon where it prefers limestone ledges and peaks.

Pack rats use cholla joints to build impenetrable thickets around their nests, affording them protection from predators such as coyote and fox. Historic pack rat middens indicate that this cholla is a relatively recent arrival, having occupied much of its present range for only about 14,000 years—about the time since the last glacial retreat.

Desert Larkspur *Delphinium parishii* —Photograph © 2006 by Kate Watters

Desert Larkspur *Delphinium parishii*
—Photograph © 2006 by Glenn Rink

Teddy-Bear Cholla *Cylindropuntia bigelovii* var. *bigelovii* —Photograph © 2006 by Wendy Hodgson

Whipple Cholla
CACTUS FAMILY (Cactaceae)

Cylindropuntia whipplei
(*Opuntia whipplei*)

Plants: Cylindrical, jointed, succulent shrub, mostly 10 to 60 centimeters tall, sometimes to 1.3 meters tall. **Spines:** Central and lateral, three to ten per areole, usually four central spines forming a cross, to 5 centimeters long. **Flowers:** Yellow to greenish yellow, 1.5 to 3 centimeters long. **Fruit:** Yellow to green, fleshy, spineless berry. **Flowering Season:** May to July. **Elevation:** 2,800 to 7,600 feet.

Whipple cholla, a low-growing, shrubby cactus, has striking, waxy yellow blooms for a few months during the year. The blooms mature into spineless yellow fruits that persist well into the fall. It occurs throughout the northern half of Arizona and into Nevada, Utah, Colorado, and New Mexico. In the Grand Canyon, it grows on both rims, extending down to river level, where it is most common in Marble Canyon. It is one of the few frost-tolerant chollas, and it grows in deep soils of slopes, flats, and dry side canyons.

Another common species, **buckhorn cholla (*Cylindropuntia acanthocarpa*)**, is typically taller, has thicker stems, and its fruits are dry at maturity. While overlapping in range, you will primarily find buckhorn cholla in warmer reaches of the Grand Canyon below Phantom Ranch, with its range extending into Mexico.

Native people eat Whipple cholla fruits fresh, boiled, or dried and ground into a flour. Medicinal tea is derived from roots. Whipple cholla honors Lieutenant Amiel Weeks Whipple (1817–1863) who led the 1853 Pacific Railroad survey from the Mississippi River to the Pacific Ocean.

Many-Headed Barrel Cactus
CACTUS FAMILY (Cactaceae)

Echinocactus polycephalus
var. *xeranthemoides*

Plants: Solitary or clustered, succulent perennial, with ribs, to 50 centimeters tall. **Spines:** Central spines flattened, straight to curved, one to four; radial spines six to fourteen. **Flowers:** Yellow, cottony below, 4 to 6 centimeters long. **Fruit:** Densely hairy, dry berry. **Flowering Season:** July to August. **Elevation:** 1,400 to 5,600 feet.

Many-headed barrel cactus can grow singularly or in clumps of up to thirty stems, almost 2 meters wide. This large, statuesque cactus, also called *cotton-top cactus*, begins to produce multiple heads after twenty years, living as long as one hundred years. The cottony yellow flowers do not fully open because they are wedged between the hardy spines. Plants occur in areas with regular frost, often growing on warmer, south-facing slopes and ledges to offset the harmful effects of cold temperatures.

Many-headed barrel cactus is commonly confused with California barrel cactus (*Ferocactus cylindraceus*), but can be distinguished by its many stems and the white cottony fibers on the flowers and fruits. Also, the central spines of California barrel cactus are slightly longer than the radial spines, which look like whiskers. Compare these to the radial and central spines of multi-headed barrel cactus, which are frequently the same size and difficult to distinguish.

Many-headed barrel cactus is restricted to the southernmost limit of the Great Basin desert extending into the Mojave Desert. In the Grand Canyon region, plants occur from the Vermillion Cliffs to Lake Mead, and along the Little Colorado River upstream to Grand Falls.

Earlier botanical studies listed the mostly Sonoran Desert variety of multi-headed barrel cactus (*Echinocactus polycephalus* var. *polycephalus*) as occurring in the western part of the Grand Canyon. Recent studies reveal that all Grand Canyon plants belong to variety *xeranthemoides*. Late Pleistocene pack rat middens suggest that both varieties occurred within their present ranges for at least 30,000 years.

Animals gather the fruits and aid in seed dispersal, which is further facilitated by a natural hole at the base of the detached fruits. Havasupai people ate the seeds fresh, or dried and ground them to make porridge. Others used the curved, thick spines as awls in basket making.

Whipple Cholla *Cylindropuntia whipplei*
—Photograph © 2006 by Kate Thompson

Buckhorn Cholla *Cylindropuntia acanthocarpa*
—Photograph © 2006 by Kate Watters

Many-Headed Barrel Cactus *Echinocactus polycephalus*
var. *xeranthemoides* —Photograph © 2006 by Geoff Gourley

Engelmann Hedgehog   *Echinocereus engelmannii*
CACTUS FAMILY (Cactaceae)

Plants: Cylindrical, succulent perennial, without prominent tubercles, to 45 centimeters tall. **Spines:** Central spines straight, two to seven, to 7.5 centimeters long; radial spines less prominent, six to fourteen, to 1.2 centimeters long. **Flowers:** Magenta to purple, to 7.5 centimeters long. **Fruit:** Red pulpy berry. **Flowering Season:** February to June. **Elevation:** 1,200 to 5,700 feet.

During the spring, Engelmann hedgehog dapples the landscape with clusters of magenta and purple blossoms, blooming at the same time as its red-flowered cousin, claret-cup cactus (*Echinocereus triglochidiatus*). In clumps of up to sixty ribbed stems, Engelmann hedgehog lacks the prominent swellings that claret-cup cactus has on its stems. Early botanists Elzada Clover and Lois Jotter Cutter noted these smaller, cliff-hugging forms of hedgehog cactus and attributed their size to lack of soil and extreme heat. Later botanists realized these cacti were a separate speces and named them *E. engelmannii*.

Engelmann hedgehog occurs from California, Nevada, Utah, Arizona, and into Mexico. In the Grand Canyon, it grows from Lees Ferry to Grand Wash Cliffs, extending to the rims. It forms clumps on rock ledges high above the river, on steep talus slopes, limestone ledges, and boulder-strewn deltas of side canyons.

Unlike the sturdy petals of claret-cup cactus, the thin, delicate petals of Engelmann hedgehog allow only for pollination by bees and other small insects. Most flowers in the cactus family are highly efficient. If on the first day a plant receives enough visits by pollinators to fertilize its ovaries, the flowers close that evening and may not open for a second day. The pleasant-tasting fruits are rich in sugar, and the small, black seeds are high in fat, making them an attractive food for birds and mammals. The specific epithet honors George Engelmann (1809–1884), botanist and describer of many plant species in the western United States.

Claret-Cup Cactus   *Echinocereus triglochidiatus*
CACTUS FAMILY (Cactaceae)

Plants: Cylindrical, succulent perennial, with tubercles, to 70 centimeters tall. **Spines:** Angled at base, central spines straight, zero to six, to 12 centimeters long; radial spines, up to twenty-five, difficult to distinguish from central. **Flowers:** Bright orange to dark red, to 10 centimeters long. **Fruit:** Yellowish green to green, fleshy berry. **Flowering Season:** March to June. **Elevation:** 1,300 to 7,900 feet.

Clusters of claret-cup cacti embellish slopes and ledges with their fiery red blossoms. The long-lasting, sturdy, tubular flowers separate claret-cup cacti from other members of this genus. Also distinguishing them are their ribbed stems divided into swellings called *tubercles*, which bear areoles with white feltlike hairs.

Claret-cup cactus grows in a variety of habitats including desertscrub and coniferous forest from California east to Colorado and Texas and south into Mexico. Plants occur throughout the Mojave and Sonoran Deserts. In the Grand Canyon, plants inhabit rocky slopes, ledges, and side canyons the length of the river.

Another Grand Canyon claret-cup cactus (***Echinocereus coccineus***) grows in the warmer, western end of the Grand Canyon, known from only a few sites. These plants form huge, rounded clumps with up to five hundred stems bearing crimson or scarlet red flowers. In these claret-cup cacti, male and female reproductive organs may be on separate plants, an anomaly in the cactus family. To distinguish between the two species, look for the curved and twisted spines of *E. coccineus*, rounded at the base, surrounding one distinct central spine. When not in flower, both claret-cup cacti may be distinguished from the related Engelmann hedgehog (*E. engelmannii*) by their larger clumps and shorter stems and spines, giving them a greener appearance.

The tubular red flowers of both claret cups remain open for a number of days to attract hummingbirds that perch on the sturdy, waxy petals to gather the flower's nectar. Through capillary action, specialized flower parts allow the nectar to move upwards against gravity towards the top of the flower, so the pollinator will not damage the young ovary that rests at the base. As the juicy fruits mature, their spines fall off, making the sweet, white pulp more appealing to birds and mammals.

Engelmann Hedgehog *Echinocereus engelmannii* —Photograph © 2006 by Lisa A. Hahn

Claret-Cup Cactus *Echinocereus triglochidiatus* —Photograph © 2006 by Glenn Rink

California Barrel Cactus   *Ferocactus cylindraceus*
CACTUS FAMILY (Cactaceae)   (*Ferocactus acanthodes*)

Plants: Cylindrical, succulent perennial, with ribs, to 3 meters tall. **Spines:** Central spines flattened, straight to slightly curved, four; radial spines slender, fifteen to twenty-five. **Flowers:** At apex of plant, yellow, sometimes tinged with red, to 6 centimeters long. **Fruit:** Dry berry with fringed scales. **Flowering Season:** April to June. **Elevation:** 1,200 to 3,000 feet.

Barrel cacti are exquisitely adapted to arid regions. Their succulent stems store considerable amounts of water and the ribs enable them to swell and then shrink during dry periods. Barrel cacti minimize water loss by opening the pores on their surface during cooler evening temperatures. Their shallow roots allow for rapid uptake of sporadic rainfall, but only lightly anchor them in the soil. The sheer weight of their stout bodies can cause them to fall from precipitous cliffs. Seedlings grow slowly and one-year-old plants may be only 1 centimeter tall! Throughout its range, California barrel cactus usually has golden yellow spines, but Grand Canyon plants display spines varying from yellow to pink to red.

California barrel cactus occurs from California, Utah, and Arizona, south to mainland Mexico and Baja California. In the Grand Canyon, it grows on rocky, open slopes and cliffs from around Serpentine Canyon downstream to Grand Wash Cliffs. In the late 1930s, Elzada Clover noted that as the water rose in Lake Mead, "many of these large plants were submerged and floating, while others were entirely covered except for an inch or two of bristling spines."

While bees are the primary pollinators, ants are also attracted to the nectar of this plant's flowers. During drought, bighorn sheep survive on water-rich stems of barrel cactus. Some indigenous people also macerated the pulp to extract the juice, but it reportedly tastes bad. Hualapai people use slices of stem pulp to provide the steam in the agave roasting process.

Graham's Fishhook Cactus   *Mammillaria grahamii* var. *grahamii*
CACTUS FAMILY (Cactaceae)   (*M. microcarpa*)

Plants: Cylindrical, succulent perennial, to 15 centimeters tall. **Spines:** Central spines reddish brown to black, hooked, one to three; radial spines whitish to brown-tipped, straight, seventeen to thirty-five. **Flowers:** In ringlets at the apex, bright rose pink to rose purple, to 4.5 centimeters long. **Fruit:** Club-shaped berry, red at maturity. **Flowering Season:** April to September. **Elevation:** 1,200 to 5,000 feet.

A crown of dainty, bright rose pink flowers burst from the top of the spiny stem of Graham's fishhook cactus, also called *pincushion cactus*. Flowers blossom for only a few days, and plants can produce two or more cycles of flowers in one season. Nurse plants, such as catclaw acacia (*Acacia greggii*) and cholla (*Cylindropuntia* species), along with this plant's many spines, protect the plant from heat, cold, and herbivores.

Graham's fishhook cactus occurs in warm desertscrub communities from California east to Texas and south into northern Mexico. In the Grand Canyon, it grows in sandy or gravelly soils in side canyons, washes, desert flats, slopes, and debris flows throughout the length of the river.

A related and very similar species, **corkseed pincushion cactus (***Mammallaria tetrancistra***)** is less common in the Grand Canyon. Graham's fishhook has seventeen to thirty-five radial spines, while corkseed pincushion cactus has thirty to sixty. As the common name suggests, corkseed pincushion cactus has a light-colored corky appendage on the black seeds, which is lacking in Graham's fishhook cactus.

California Barrel Cactus *Ferocactus cylindraceus*
—Photograph © 2006 by Lisa A. Hahn

Flowers of Graham's Fishhook Cactus *Mammallaria grahamii* var. *grahamii* —Photograph © 2006 by David Edwards

Fruit of Graham's Fishhook Cactus *Mammallaria grahamii* var. *grahamii*
—Photograph © 2006 by Kristin Huisinga

PRICKLY-PEAR　　　　　　　　　　　　　　*Opuntia* species
CACTUS FAMILY (Cactaceae)

Even cactus experts struggle to discern which prickly-pear is which because a number of Grand Canyon populations appear to hybridize. They have a unique adaptation to maximize pollen transfer: the anthers squeeze inward when triggered by pollinators, which places pollen directly on their bodies.

The crimson red fruits of many prickly-pear, called *tunas*, can be eaten fresh or mashed and strained for juices, jellies, and syrups. People collect the tender young pads and remove the spines and *glochids*, which are minute needlelike spinelets that arise from areoles. The pads, which are eaten raw, boiled, or fried, are calcium-rich, and the slimy mucilage aids diabetics with sugar control. In the Grand Canyon, some plants have exceptionally large pads, which often grow in association with archeological sites. Indigenous people may have imported these plants from the south and cultivated them for their large pads and fruits.

White fuzz on the pads and fruits of prickly-pears indicates the presence of a cochineal-scale insect (*Dactylopius coccus*). The female secretes a weblike, wax-based material to prevent desiccation and to camouflage herself. She also manufactures carminic acid, a deep maroon pigment, to repel potential predators such as ants. Because this pigment was long-lasting and brighter than other red pigments in the Old World, people in pre-Columbian Mexico farmed cochineal by planting pads already inoculated with the scale insect. The Spaniards were impressed with this scarlet color and shipped dyed textiles to Europe where they became a prized commodity. Cochineal cultivation continues today, primarily in Mexico.

Grand Canyon Beavertail Cactus　　　*Opuntia basilaris* var. *longiareolata*
CACTUS FAMILY (Cactaceae)

Plants: Flat-stemmed, succulent perennial, to 2 meters tall. **Pads:** Spatulate, blue green tinged with maroon, 9 to 25 centimeters long. **Spines:** Lacking, with many glochids per areole. **Flowers:** Bright pink to magenta, 2.5 to 4 centimeters long. **Fruit:** Fleshy berry. **Flowering Season:** March to May. **Elevation:** 1,200 to 6,800 feet.

Beavertail is unique from all other prickly-pears because it lacks spines. Like all prickly-pears, it has numerous glochids. If you brush against or touch the pads, the glochids detach by the hundreds, effectively deterring harvest or grazing.

Grand Canyon beavertail cactus is restricted to northern Arizona and southern Utah, but it boasts a tremendous elevational range, growing from dry deserts to high mountains. In the Grand Canyon, it occurs throughout the Inner Canyon from Lees Ferry to Grand Wash Cliffs and is common on the Tonto Plateau. Two varieties overlap but mostly occupy separate portions of the Grand Canyon. Dominant in the eastern and central parts is variety *longiareolata*, which has spatula-shaped pads and few, widely spaced areoles per pad. Most common in the western end is variety *basilaris*. Its pads more strongly resemble the obovate shape of a beaver's tail, with numerous, more closely spaced areoles.

Pack rats and other desert dwellers savor the seeds and have learned how to avoid the glochids while gathering the fruits. Native people fashion efficient whisk brooms from bundles of snakeweed (*Gutierrezia sarothrae*) to brush off the glochids and make pads and fruits safe to handle and eat. This species was named by Elzada Clover and Lois Jotter Cutter, who first collected it in the Grand Canyon.

Engelmann Prickly-Pear, Nopal   *Opuntia engelmannii* var. *engelmannii*
CACTUS FAMILY (Cactaceae)

Plants: Flat-stemmed, succulent perennial, to 2 meters tall. **Pads:** Circular to obovate, 20 to 40 centimeters long. **Spines:** Zero to five per areole, evenly distributed, white to yellow, aging to gray, dark at base, longer spines bent backward. **Flowers:** Uniform in color; clear yellow, pink, or red; 3 to 4 centimeters long. **Fruit:** Deep red to purple, fleshy berry. **Flowering Season:** April to May. **Elevation:** 1,200 to 7,900 feet.

Flowers of a single Engelmann prickly-pear are a uniform color, but the colors in the species range from clear yellow to pink to red. Engelmann prickly-pear occurs from California and Nevada, east to New Mexico and Texas, and south into Mexico. Plants are generally absent on the Colorado Plateau except for populations within the Grand Canyon, where it is common throughout the Inner Canyon.

Grand Canyon Beavertail Cactus
Opuntia basilaris var. *longiareolata*
—Photograph © 2006 by Kyle George

Engelmann Prickly-Pear *Opuntia engelmannii* var. *engelmannii* —Photograph © 2006 by Lisa A. Hahn

Engelmann Prickly-Pear *Opuntia engelmannii* var. *engelmannii* —Photograph © 2006 by David Edwards

A related species, **pancake prickly-pear (*Opuntia chlorotica*)** is a treelike cactus with a stout trunk and large, bluish green, circular pads bearing yellow spines. Flowers are yellow, sometimes tinged with red and dense glochids engulf both the flower cup and fruit. Compare the smaller, drier, barrel-shaped fruits of pancake prickly-pear with those of Engelmann prickly-pear, which are fewer-spined, large, juicy, and dark purple throughout.

Brown-Spined Prickly-Pear *Opuntia phaeacantha*
CACTUS FAMILY (Cactaceae)

Plants: Low, succulent perennial, to 1 meter tall. **Pads:** Obovate to circular, 10 to 25 centimeters long. **Spines:** Zero to eight per areole; central spines limited to the top portion of pad, from red brown to black; shorter spines whitish. **Flowers:** Yellow with red at the base, rarely pink to red, 3 to 4 centimeters long. **Fruit:** Wine red to purple, fleshy berry, sometimes juicy with age. **Flowering Season:** April to June. **Elevation:** 2,000 to 6,900 feet.

Brown-spined prickly-pear closely resembles its relative Engelmann prickly-pear (*Opuntia engelmannii* var. *engelmannii*). However, as its name suggests, many of its spines are brown, especially on the upper part of each pad. It forms large clumps that trail along the ground and displays yellow or pink flowers tinged with red at the base.

Brown-spined prickly-pear occurs from California to Arizona, Utah, Colorado, and as far east as Tennessee. In the Grand Canyon, plants grow in sandy and rocky substrates on flats and slopes throughout the Inner Canyon and up to the rims. It is particularly common along the Tonto Plateau, where it often hybridizes with grizzly-bear prickly-pear cactus (*O. polyacantha* var. *erinacea*).

Grizzly-Bear Prickly-Pear *Opuntia polyacantha* var. *erinacea*
CACTUS FAMILY (Cactaceae)

Plants: Flat-stemmed, succulent perennial, to 3 meters tall. **Pads:** Elliptic to obovate, green to greenish gray, to 20 centimeters long. **Spines:** one to eighteen, somewhat flexible. **Flowers:** Yellow to pink or magenta, 2.5 to 4 centimeters long. **Fruit:** Fleshy berry, becoming dry. **Flowering Season:** April to July. **Elevation:** 1,200 to 7,900 feet.

Grizzly-bear prickly-pear cactus has densely spiny pads that burst with decadent yellow or magenta blossoms during spring and summer. The Grand Canyon hosts three of the four varieties of this species, which can be differentiated by the arrangement and character of their spines. **Porcupine prickly-pear cactus (*Opuntia polyacantha* var. *hystricina*)** may remind you of a cartoon figure with spiked hair, with its long, stiff, and straight spines that stick up, especially along the terminal edges of the pads. Endemic to the Grand Canyon area, **Navajo Bridge prickly-pear (*O. p.* var. *nicholii*)** occurs primarily near its namesake in Marble Canyon. It has very long, thick spines and dark, maroon flowers. Compare these to the shaggy, threadlike spines of grizzly-bear prickly-pear.

Grizzly-bear prickly-pear cactus occurs in California, Nevada, Arizona, and Utah. In the Grand Canyon, it grows throughout the Inner Canyon and is especially abundant on the Tonto Plateau.

The early-maturing, fleshy fruits are stewed with meat, roasted in hot coals, or eaten raw. As a poultice, the flesh is used for treating skin sores and infections. The stems are also used to fix color on hides, and the spines are used for fishhooks.

Pancake Prickly-Pear *Opuntia chlorotica*
—Photograph © 2006 by Kate Watters

Brown-Spined Prickly-Pear *Opuntia phaeacantha*
—Photograph © 2006 by David Inouye

Grizzly-Bear Prickly-Pear
Opuntia polyacantha var. *erinacea*
—Photograph © 2006 by Glenn Rink

Porcupine Prickly-Pear
Opuntia polyacantha var. *hystricina*
—Photograph © 2006 by Lisa A. Hahn

Creosotebush
CALTROP FAMILY (Zygophyllaceae)

Larrea tridentata var. *tridentata*
(*Larrea divaricata*)

Plants: Many-stemmed, evergreen shrub, to 4 meters tall. **Leaves:** Paired, curved lanceolate, 5 to 10 millimeters long. **Flowers:** Solitary, yellow, five-petaled, to 2.5 centimeters long. **Fruit:** Furry capsule. **Flowering Season:** April to May. **Elevation:** 1,200 to 4,000 feet

For some, the noteworthy scent of creosotebush after a rainstorm evokes the very essence of the desert. It is among the most drought-tolerant plants in North America and is one of the longest-lived plants on the planet. One individual in the Mojave Desert may be nearly 11,000 years old. Ecologically creosotebush is an important nurse plant, providing protection for other plants from extreme temperatures, wind, and herbivory. In the soil beneath creosotebush, many small mammals build burrows, which other reptiles and invertebrates may inhabit at a later time. The bush hosts over sixty species of insects, with nearly one-third of these depending solely on its flowers for survival. Very few creatures can tolerate the potent, resin-laden leaves, but jackrabbits (*Lepus* species) feed on them in times of drought.

Although scarce in the cold Great Basin Desert, creosotebush is one of the most common and widespread shrubs in the other three warm North American deserts. In the Grand Canyon, it grows on sandy and gravelly desert slopes, from River Mile 167 downstream to Grand Wash Cliffs.

Native people in the Southwest and Mexico consider creosotebush one of their most essential plants because it offers a myriad of goods including medicinal teas and salves, wood, and building material. The sticky secretions of lac insects (*Tachardiella larreae*) that live on the stems are used as a sealant and glue. Creosote oil, a common wood preservative, is a petroleum product and is *not* derived from creosotebush.

Wide-Wing Spring-Parsley
CARROT FAMILY (Apiaceae/Umbelliferae)

Cymopterus purpurascens

Plants: Low-growing perennial herb, 3 to 15 centimeters tall. **Leaves:** Once or twice pinnately lobed, with petioles, to 9 centimeters long. **Flowers:** In terminal umbels, purplish, to 2.5 centimeters long. **Fruit:** Winged. **Flowering Season:** April to May. **Elevation:** 4,500 to 7,000 feet

The highly divided leaves of wide-wing spring-parsley hug the ground, and just above the mat of leaves lies a bundle of deep purple flowers. Appearing in early spring, these blossoms later mature into papery, winged, dry fruits.

Wide-wing spring-parsley occurs from California, Arizona, and New Mexico, north to Colorado and Idaho. In the Grand Canyon, plants grow on the Tonto Plateau, at higher elevations along trails, and in open flats often among sagebrush and piñon-juniper communities.

While many members of this family are highly poisonous, the root of wide-wing spring-parsley offers a reliable spring food, much like a carrot, that native people harvest in spring prior to flowering.

Creosotebush *Larrea tridentata* var. *tridentata*
—Photograph © 2006 by Celia Southwick

Creosotebush *Larrea tridentata* var. *tridentata*
—Illustration © 2006 by Lisa Kearsley

Wide-Wing Spring-Parsley
Cymopterus purpurascens
—Photograph © 2006 by Celia Southwick

Turpentine-Broom
Thamnosma montana
CITRUS FAMILY (Rutaceae)

Plants: Compact shrub, 10 to 60 centimeters tall. **Leaves:** Linear, to 1.5 centimeters long. **Flowers:** In racemes, purple, four petals fused at base, 8 to 12 millimeters long. **Fruit:** Capsule with two spherical globes. **Flowering Season:** February to May. **Elevation:** 1,200 to 5,900 feet.

Turpentine-broom heralds spring with indigo purple blooms clinging to grayish green stems that yellow by summer. The reduced leaves drop in summer, and then the plant appears spiny. At this stage, you may confuse it with Mormon tea (*Ephedra* species) or rabbitbrush (*Ericameria nauseosus*), but one sniff should identify turpentine-broom because it emits a pungent odor. Turpentine-broom is one of two plants in the citrus family found in the Grand Canyon. The other is pale hoptree (*Ptelea trifoliata* ssp. *pallida*). The stems, leaves, and fruit of plants in both genera are covered with sunken glands that release a strong aromatic fragrance when bruised, much like the rinds and leaves of their citrus relatives.

Turpentine-broom occurs from California east to Arizona and Utah and into Mexico. In the Grand Canyon, it most commonly grows on dry mesas and slopes throughout the Inner Canyon from the base of the Redwall Limestone to the river.

Turpentine-broom is a host plant for some swallowtail butterflies (*Papilio* species). Female butterflies recognize turpentine-broom by smell, sensing the gland-laden leaves and stems with their feet. They lay their eggs on the plants, and the hefty caterpillars eat only turpentine-broom as they mature. Predators dislike their taste and readily avoid these caterpillars. The genus name, *Thamnosma*, comes from the Greek words *thamnos*, meaning "shrub," and *osme*, meaning "odorous."

Dogbane
Apocynum cannabinum
DOGBANE FAMILY (Apocynaceae)

Plants: Perennial herb, to 1.2 meters tall. **Leaves:** Opposite, oblanceolate, 5 to 8 centimeters. **Flowers:** In many-flowered cymes, white, urn-shaped, tiny. **Fruit:** Podlike follicles. **Flowering Season:** May to July. **Elevation:** 1,500 to 6,500 feet.

Dogbane grows abundantly near the river's edge and has deep red stems that branch near the top. The thick, shiny leaves and broken stems exude a milky, latexlike sap. The long, slender fruit resembles a green bean pod and opens on one side at maturity to unleash a throng of brown, silky-haired seeds, much like the fruits of the closely related milkweed family.

Dogbane has a broad range, extending from California to British Columbia and into the eastern United States. In the Grand Canyon, it grows in moist side canyons, springs, and riparian areas the entire length of the river, extending up to the rims.

The small, sweetly fragrant flowers attract bees, which yield a high-grade honey. The larvae of the metallic, blue green dogbane leaf beetle (*Chrysochus auratus*) tunnel through the soil to feed on the roots, and adults ingest the leaves, becoming poisonous to predators. As its other common name, *Indian hemp*, implies, this plant provided people with durable fiber, which was woven into twine and has a breaking point of several hundred pounds. Dried stems were harvested, flattened, and then split to remove the inner core, leaving the fiber to roll into twine. Many herbalists warn of dogbane's potency because it increases perspiration, acts as a laxative, and is a strong expectorant. It also contains cardiac glycosides, which have been used for centuries to treat heart problems and to concoct poison.

Turpentine-Broom
Thamnosma montana
—Photograph © 2006
by Raechel Running
RMRfotoarts.com

Turpentine-Broom
Thamnosma montana
—Photograph © 2006
by Celia Southwick

Dogbane
Apocynum cannabinum
—Photograph © 2006
by David Edwards

EPHEDRA, MORMON TEA *Ephedra* species
EPHEDRA FAMILY (Ephedraceae)

Ephedras are gymnosperms like pines and junipers and have separate male and female plants but are more primitive. The "flowers" visited by insects in the spring are actually male or female cones. Cone-bearing plants do not produce food for their embryos, but ephedra appears to make an attempt to do this thru fusion between a second sperm and the embryonic sac. Some botanists speculate that ephedra may be the "missing link" between the cone-bearing gymnosperms and more-evolved, flower-bearing angiosperms. Ephedras share a common ancestor with Welwitschia (*Welwitschia* species), a sprawling, octopus-shaped plant from the Namib Desert in Africa, which bears similar cones.

Mojave Ephedra, Mormon Tea *Ephedra fasciculata*

Plants: Pale green, perennial shrub, with opposite branches, to 1 meter tall. **Leaves:** Opposite, two per node, with persistent white sheath, 1 to 3 millimeters long. **Cones:** Male cones two to several; female cones sessile or short-stalked, with unclawed scales. **Seed:** Furrowed, pale brown, solitary.
Flowering Season: March to May. **Elevation:** 1,200 to 4,000 feet.

Torrey Mormon Tea *Ephedra torreyana*

Plants: Blue green, perennial shrub, with whorled branches, to 1 meter tall. **Leaves:** Whorled, three per node, 2 to 5 millimeters long. **Cones:** Male cones sessile, whorled, one to four; female cones sessile, with clawed, translucent scales. **Seed:** Rough, pale brown to yellow green, solitary or paired.
Flowering Season: April to May. **Elevation:** 1,200 to 5,200 feet.

These two species of ephedra in the Grand Canyon appear very similar. Mojave ephedra is usually prostrate and pale green, turning yellow with age, compared to the erect, blue-green nature of Torrey Mormon tea. However, you need more than color differences to separate the two species. Look closely at their minute leaves and cones. In Mojave ephedra, there are two leaves per node and the female cone scales are unclawed, while Torrey Mormon tea has three leaves per node and the female cones bear translucent claws.

Over sixty species of ephedra inhabit temperate and warm regions worldwide. These two ephedra are the most common species found on rocky beaches, talus slopes, and mesas the length of the Grand Canyon. Torrey Mormon tea is most common in the blackbrush communities of the Tonto Plateau.

Another species, **Nevada Mormon tea (*Ephedra nevadensis*)** grows in the western Grand Canyon near Toroweap. While rare at river level, the higher elevation **green ephedra (*E. viridis*)** is more robust and a deeper, vivid shade of green. **Cutler's ephedra (*E. cutleri*)** has the most restricted distribution, only recorded from Havasu Canyon.

Several related *Ephedra* species, all commonly called *Chinese ephedra* or *ma huang*, have been marketed for weight loss, but their misuse have caused death. Ephedra plants produce caffeine and ephedrine, alkaloids with stimulant properties that have been used as decongestants to relieve colds and alleviate asthma. Green ephedra branches can be made into a tea, from which the common name is derived. You might see bighorn sheep eating this plant, especially during winter.

Males cones of Ephedra *Ephedra* species
—Photograph © 2006 by Kathy Darrow

Female cones of Ephedra *Ephedra* species
—Photograph © 2006 by David Edwards

Ephedra *Ephedra* species —Photograph © 2006 by Celia Southwick

Suncup
Camissonia multijuga
EVENING-PRIMROSE FAMILY (Onagraceae)

Plants: Annual or biennial herb, 10 to 50 centimeters tall. **Leaves:** In basal rosettes, highly dissected, purple dotted, 4 to 30 centimeters long. **Flowers:** In racemes, yellow, 5 to 13 millimeters long. **Fruit:** Slender capsule. **Flowering Season:** April to June. **Elevation:** 1,200 to 4,200 feet.

Suncups are the denizens of cobbled beaches and washes, opening their sunny flowers in the early morning and closing them during the heat of the day. You could confuse them with members of the mustard family because they have four petals and similarly shaped fruit. Unlike mustards, suncups have an inferior ovary located at the base of a long floral tube. Suncups have four sepals and four petals that protect the flower in bud. Like others in the evening-primrose family, the sepals dramatically bend backward when the flower is in full bloom.

Suncup grows throughout creosotebush communities in Nevada, Utah, and Arizona. In the Grand Canyon, it is the most common species in this genus and grows on rocky beaches and side canyons the length of the river. However, be warned as there are nearly ten different species of suncup that speckle the slopes of the Inner Canyon, especially during wet years. Two of them are rare perennial subspecies of **Kaibab suncup (*Camissonia specuicola*)** that are endemic to the Grand Canyon area.

While feeding on pollen grains, solitary bees inadvertently pollinate the small, yellow flowers. With their long proboscises, hawk moths gather nectar, often dusting their breasts with pollen and transferring it to another flower. The genus name, *Camissonia*, commemorates A. L. Von Chamisso (1781–1838), a German poet, naturalist, and botanist on the ship *Rurik* that visited California in 1816.

Hooker's Evening-Primrose
Oenothera elata ssp. *hookeri*
EVENING-PRIMROSE FAMILY (Onagraceae)
(*O. hookeri*)

Plants: Biennial herb, to 2 meters tall. **Leaves:** Alternate or basal, lanceolate, 3 to 30 centimeters long. **Flowers:** In spikes, yellow. **Fruit:** Erect capsule. **Flowering Season:** July to October. **Elevation:** 1,900 to 3,100 feet.

Hooker's evening-primrose produces a basal rosette of leaves with distinct white midveins during its first year. During its second year of life, it sends up a handsome flowering stalk. The tall, lanky stalks remain long after flowering, dispersing their tiny seeds from slender, four-chambered capsules. You will likely see Hooker's evening-primrose in one of its growth stages as you hike along Saddle, Stone, or Deer Creeks.

Hooker's evening-primrose grows from Washington to Montana and south to Panama. In the Grand Canyon, it occurs along the river's edge and in side streams where its boastful blooms line waterways in late summer. It grows from Lees Ferry to Deer Creek, but likely occurs farther downstream as well.

Resembling Hooker's evening-primrose in stature, **longstem evening-primrose (*Oenothera longissima*)** has a longer floral tube (6 to 12 centimeters long compared to Hooker's 2.5 to 5 centimeters). It resides on the North Rim and at scattered localities throughout the Inner Canyon.

Hooker's evening-primrose honors Sir Joseph Dalton Hooker (1817–1911), a nineteenth-century botanist and director of the Royal Botanical Gardens at Kew in Britain. He traveled throughout the world in search of plants and was a friend of Charles Darwin and John Muir.

Suncup *Camissonia multijuga* —Photograph © 2006 by Tyler Williams

Hooker's Evening-Primrose
Oenothera elata ssp. *hookeri*
—Photograph © 2006 by Kristin Huisinga

Basal leaves of Suncup *Camissonia multijuga*
—Photograph © 2006 by Tyler Williams

Pale Evening-Primrose
Oenothera pallida
EVENING-PRIMROSE FAMILY (Onagraceae)

Plants: Annual, biennial, or perennial herb, 10 to 70 centimeters tall. **Leaves:** Alternate, linear to lanceolate, dark green, minutely lobed, 1 to 8 centimeters long. **Flowers:** Solitary in leaf axils, white, 1 to 3.5 centimeters long. **Fruit:** Curved capsule. **Flowering Season:** March to October. **Elevation:** 1,450 to 3,600 feet.

The sweetly fragrant, delicate blossoms of pale evening-primrose have four petals and backward-bending sepals. In the evening as the light fades, the strong scent of the flowers attracts hawk moths (*Manduca* species), which hone in on the papery, moon-colored blossoms. Often mistaken for a flower stalk, the very long flower tubes house sucrose-rich nectar. Hawk moths hover above the flowers and unravel their 10-centimeter-long proboscises to drink this liquid energy, getting dusted with pollen in the process.

Pale evening-primrose occurs throughout the West except in California. In the Grand Canyon, it grows on beaches and sand dunes along the river from Lees Ferry to Granite Park. Eleven other evening-primrose species grow within Grand Canyon National Park.

Tufted evening-primrose (*Oenothera caespitosa*) is low-growing, not upright and shrubby like pale evening-primrose. Its four-petaled white flowers are larger (to 20 centimeters long), and its longer leaves (to 20 centimeters) bear shaggy hairs along the margins. While overlapping in range with pale evening-primrose, tufted evening-primrose extends to higher elevations on both rims. Sand-verbena (*Abronia elliptica*), another beach dweller with a lovely scent similar to that of pale evening-primrose, has very different flowers.

Sand-Verbena
Abronia elliptica
FOUR O'CLOCK FAMILY (Nyctaginaceae)

Plants: Creeping annual or biennial herb, 10 to 50 centimeters long. **Leaves:** Opposite, oval to elliptic, 1 to 5 centimeters long, glandular-hairy. **Flowers:** In headlike clusters, enclosed from below by bracts; tubular white sepals that are up to 5 centimeters long; no petals. **Fruit:** Papery-winged, one-seeded. **Flowering Season:** April to November. **Elevation:** 1,375 to 3,000 feet.

When in flower, sand-verbena's sweet aroma permeates the air, beginning at dusk and lasting through the night. This fragrance provides the initial cue for night-flying pollinators such as hawk moths in the family Sphingidae. The bright white flowers guide them to their reward. The elliptical, glandular leaves of this trailing plant attract sand, giving the entire plant a grainy, coarse appearance. These hairs and sand provide a reflective surface, which slows evaporative water loss.

Sand-verbena occurs from Wyoming south to New Mexico and west to Nevada. Plants in the genus *Abronia* are only found west of the Mississippi River. In the Grand Canyon, plants grow on sandy beaches and dunes the length of the river.

Sand-verbena's white flower clusters help distinguish it from other species that share its habitat on sandy beaches, such as pale evening-primrose (*Oenothera pallida*). **Dwarf sand-verbena (*A. nana*)**, a shorter less common species, has white to purple flowers and grows at higher elevations to 6,100 feet.

The genus name, *Abronia*, is derived from the Greek word *abros* or *habros*, meaning "delicate" or "soft," referring to the flower bracts.

Tufted Evening-Primrose *Oenothera caespitosa*
—Photograph © 2006 by Kate Watters

Pale Evening-Primrose *Oenothera pallida*
—Photograph © 2006 by Joseph Bennion

Sand-Verbena *Abronia elliptica* —Photograph © 2006 by Geoff Gourley

Sand-Verbena *Abronia elliptica* —Photograph © 2006 by Kristin Huisinga

Trailing Four O'clock   *Allionia incarnata*
FOUR O'CLOCK FAMILY (Nyctaginaceae)

Plants: Trailing annual or short-lived perennial herb, to 1 meter long, glandular-hairy. **Leaves:** Opposite, oval to elliptic, to 4 centimeters long. **Flowers:** In clusters, enclosed from below by bracts; fringed, purple sepals that are 3 to 15 millimeters long. **Fruits:** One-seeded. **Flowering Season:** April to October. **Elevation:** 1,200 to 4,500 feet.

A common sight in the fall on most side canyon hikes, this delicate plant has sticky, opposite leaves. The crawling green stems and purple blooms of trailing four o'clock weave across the desert's floor. The blossom looks like a regular flower but actually contains three flowers, all of which lack petals but have brightly colored, fringed sepals.

Trailing four o'clock occurs from California east to Colorado and Texas and into South America. In the Grand Canyon, plants grow on beaches, in side canyons, and on slopes the length of the river.

The genus name, *Allionia*, honors Carlo Ludovico Allioni (1728–1804), an Italian botanist, physician, and contemporary of Carl Linnaeus.

Ringstem   *Anulocaulis leiosolenus* var. *leiosolenus*
FOUR O'CLOCK FAMILY (Nyctaginaceae)

Plants: Perennial herb, to 1.5 meters tall. **Leaves:** In a basal rosette, broadly ovate to cordate, to 15 centimeters wide. **Flowers:** Funnel-shaped, pink to white, 2 to 3 centimeters long. **Fruit:** Diamond-shaped, one-seeded. **Flowering Season:** May to June. **Elevation:** 1,700 to 4,000 feet.

Ringstem, a very unique plant, has large, leathery leaves and long tubular flowers with purple filaments that protrude up to twice the length of the flowers. The sticky bands that encircle the stems may deter ants from stealing nectar from the flowers without pollinating them.

Ringstem grows on alkaline clay and gypsum soils. The Nevada and Arizona populations are geographically isolated from those in New Mexico, Texas, and Mexico, where they grow in the Chihuahuan Desert. In the Grand Canyon, it occurs in side canyons from River Mile 18 to River Mile 175.

Desert Four O'Clock   *Mirabilis multiflora*
FOUR O'CLOCK FAMILY (Nyctaginaceae)

Plants: Spreading perennial herb. **Leaves:** Opposite, oval to cordate, 2 to 10 centimeters long. **Flowers:** In axillary clusters of three to six, brilliant magenta to purple petal-like sepals. **Fruit:** Smooth achene. **Flowering Season:** April to October. **Elevation:** 2,500 to 6,500 feet.

The eye-catching, fuchsia blooms of desert four o'clock open in the late afternoon, at the same time as the white blossoms of sand-verbena (*Abronia elliptica*) and purplish blossoms of trailing four o'clock (*Allionia incarnata*), two other members of the four o'clock family. You can distinguish desert four o'clock by its rounded, shrubby habit and distinctly larger, oval leaves.

Desert four o'clock grows in creosotebush, blackbrush, and piñon-juniper communities from California and Utah east to Texas and south to Mexico. In the Grand Canyon, it grows in dry soils on slopes and desert flats the length of the river.

The robust root of desert four o'clock is used as a poultice for wounds and to relieve stomachaches and hunger. The root also has mildly euphoric properties. The genus name, *Mirabilis*, is Latin for "marvelous" and "extraordinary," an appropriate name for this desert showstopper.

Trailing Four O'Clock *Allionia incarnata*
—Photograph © 2006 by Roger Dale

Ringstem *Anulocaulis leiosolenus* var. *leiosolenus*
—Photograph by Lori Makarick/National Park Service

Desert Four O'Clock *Mirabilis multiflora* —Photograph © 2006 by Kristin Huisinga

Arizona Centaury
GENTIAN FAMILY (Gentianaceae)

Centaurium arizonicum
(*C. calycosum* var. *arizonicum*)

Plants: Annual or biennial herb, 12 to 60 centimeters tall. **Leaves:** Opposite, lanceolate to elliptic, light green, clasping, 1 to 7 centimeters long. **Flowers:** In open cymes, star-shaped with a slender tube, pink-magenta, to 1.5 centimeters long. **Fruit:** Capsule. **Flowering Season:** March to November. **Elevation:** 1,250 to 3,000 feet.

Look for the delicate Arizona centaury along side canyon streams. The coiled, bright yellow stamens peek out from a showy tube formed of five fused, pink-magenta petals. Plants in the genus *Centaurium* are somewhat of a mystery in the botanical realm because they are little-studied, highly variable, and hybridize readily.

Arizona centaury is rare in its range from California east to Colorado and Texas and south to Mexico. It requires the permanently wet soil of springs, seeps, and perennial streams where it resides in riparian, blackbrush, and piñon-juniper communities. In the Grand Canyon, Arizona centaury flourishes on riverbanks, often growing among horsetail in marshes, and in side canyons such as Clear Creek and Three Springs.

The genus name, *Centaurium*, refers to the mythological centaur Chiron who reportedly cured himself of a poison-arrow wound with a plant from this genus. The plant's medicinal properties are useful for treating rheumatism, wounds, jaundice, and malaria.

Filaree, Storksbill
GERANIUM FAMILY (Geraniaceae)

Erodium cicutarium

Plants: Exotic annual or biennial herb, 10 to 50 centimeters long. **Leaves:** In a basal rosette, pinnately compound, highly divided, 3 to 10 centimeters long. **Flowers:** Solitary, pink to purple, five-petaled, 5 to 7 millimeters long. **Fruit:** Beak-shaped capsule. **Flowering Season:** February to August. **Elevation:** 1,200 to 7,200 feet.

In the spring, filaree blankets the ground with doilylike clusters of basal leaves and pinkish purple flowers. As the season progresses, flowers develop into beaklike fruits atop the red-tinged rosettes. The fruit exhibits a clever dispersal mechanism: As the beak dries and begins to separate, the sections curl, eventually launching seeds as if from a carnival ride. The seeds, still attached to the twisted beak sections, auger themselves into the ground and the socks of passersby.

Originally from Europe, filaree is widespread throughout North America. In the Grand Canyon, it is extremely abundant in a variety of habitats from river to rim.

A showier *Erodium* species in the Grand Canyon is native and easy to distinguish by its leaf shape alone. **Texas storksbill (*E. texanum*)** has rounded, lobed, geranium-like leaves, larger flowers, and an upright habit. Although the two species often grow together, Texas storksbill is less common.

Filaree is widely documented as an edible food source and it is suspected that the native species, Texas storksbill, was used as well. By the time serious ethnobotanical work began in the early 1900s, it is likely that people had shifted to using the already more common, exotic species, filaree.

Arizona Centaury *Centaurium arizonicum*
—Photograph © 2006 by David Edwards

Texas Storksbill *Erodium texanum*
—Photograph © 2006 by Kathy Darrow

Filaree *Erodium cicutarium* —Photograph © 2006 by Geoff Gourley

Four-Wing Saltbush   *Atriplex canescens*
GOOSEFOOT FAMILY (Chenopodiaceae)

Plants: Evergreen shrub, to 2 meters tall. **Leaves:** Alternate, linear, grayish white, to 4 centimeters long. **Flowers:** In axillary clusters; male spikelike; female elongated, with leafy bracts. **Fruit:** Four-winged seed. **Flowering Season:** March to August. **Elevation:** 2,500 to 8,000 feet.

Highly adaptable to a variety of ecosystems, four-wing saltbush requires full sunlight for growth and does not survive well in shaded areas. It usually grows in patches, with the female plants bearing clusters of green to golden fruits with four parchmentlike wings. The male plants produce pollen from dense, yellow flower clusters. Four-wing saltbush has developed a strangely clever practice of sex swapping. In response to drought or following a year of prolific seed production, some female plants may opt to switch their sex. Male plants require less water, and acting like a male affords females time to recover and amass energy because pollen is less expensive to produce than fruits.

Four-wing saltbush appears chalky green because small, white scales cover the leaves and shoots. The scales minimize evaporation during the heat of summer. The salt-storing bladders on the leaf surface give these shrubs a salty taste. In winter, these tiny bladders erupt, coating the leaves with a protective layer of salt. The roots of young seedlings partner with a fungus that facilitates nutrient and water uptake, giving the plants a competitive advantage over other seedlings that are trying to grow in nutrient-poor soils.

Four-wing saltbush grows abundantly in western North America and east to Kansas. In the Grand Canyon, it is widespread from the river to the rims.

There are six other species of *Atriplex* in the Grand Canyon, two of which grow in the river corridor. Compare the linear leaves of four-wing saltbush with the wider leaves of **shadscale (*A. confertifolia*)** and **New Mexico saltbush (*A. obovata*)**. Both are more compact shrubs with oval to spoon-shaped leaves. Shadscale is unique in having conspicuously thorny branch tips.

Shadscale *Atriplex confertifolia*
—Illustration © 2006 by Lisa Kearsley

New Mexico Saltbush
Atriplex obovata —Illustration © 2006 by Lisa Kearsley

In all seasons, four-wing saltbush leaves provide dependable forage for deer, pronghorn, elk, and rabbits. Birds and small mammals eat the seeds and take shelter under the thick branches of the shrub. The seeds are a good source of niacin and make a tasty oatmeal-style dish. In the spring, the new shoots and leaves can be eaten raw or cooked as greens. Ashes from leaves and stems enhance the color and nutritional value of blue corn products like *piiki*, a Hopi wafer bread. The genus name, *Atriplex*, is Latin for "orach," a leafy kitchen vegetable, referring to this genera's importance as a food.

Female branch of Four-Wing Saltbush
Atriplex canescens —Photograph © 2006 by Celia Southwick

Male branch of Four-Wing Saltbush
Atriplex canescens
—Photograph © 2006
by Glenn Rink

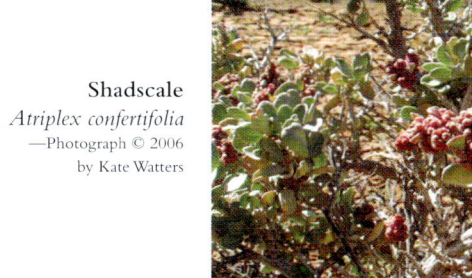

Shadscale
Atriplex confertifolia
—Photograph © 2006
by Kate Watters

Tumbleweed, Russian Thistle   *Salsola tragus*
GOOSEFOOT FAMILY (Chenopodiaceae)   (*S. iberica*)

> **Plants:** Exotic, shrubby, annual herb, to 1 meter tall. **Leaves:** Threadlike, dark green, spine-tipped, leathery, to 5 centimeters long. **Flowers:** Solitary in axils, translucent to cream-colored, to 2.5 millimeters long. **Fruit:** Winged seed. **Flowering Season:** April to September. **Elevation:** 1,200 to 7,000 feet.

Dried tumbleweed plants careening across desert highways are a common sight. When the plant matures, a specialized layer of cells in the stem facilitates the easy break between plant and root, freeing the plant to tumble in the wind and disperse seed as it travels. The highly branched skeletons of these plants can reach the size of a Volkswagen Beetle. Seed production is prolific, with up to 250,000 seeds per plant. Seeds lack a protective coat or stored food reserves. A coiled embryonic plant already equipped with chlorophyll lies within the seed, allowing it to sprout very quickly when conditions permit.

Tumbleweed is native to the arid steppes of Russia but now occurs worldwide. It is believed that Ukranian farmers unknowingly brought it to South Dakota in 1877 with imported flaxseed. By 1900, it had spread as far as the Pacific Coast. It was first reported in our region around 1900 at the Hopi Mesas in northern Arizona where it was named *pahan'uyi*, meaning "white man's plant." In the Grand Canyon, it grows in disturbed habitats from river to rim.

Bighorn sheep, pronghorn, small mammals, and domestic livestock eat the tender shoots. Small mammals and a variety of birds gorge on the seeds. Steam the edible young shoots with butter and garlic for a wild food delicacy.

Desert Seepweed, Inkweed   *Suaeda moquinii*
GOOSEFOOT FAMILY (Chenopodiaceae)   (*S. torreyana*)

> **Plants:** Perennial shrub, to 1.5 meters tall. **Leaves:** Alternate, linear to lanceolate, fleshy, flattened, to 3.5 centimeters long. **Flowers:** In axillary, leafy clusters, greenish, inconspicuous. **Fruit:** Utricle enclosed in sepals. **Flowering Season:** July to September. **Elevation:** 1,400 to 2,800 feet.

Desert seepweed forms large, conspicuous stands and is well adapted to saline environments. Its ability to store high levels of salt in its fleshy leaves increases its capacity to extract water from salty soils. Wind facilitates pollination of desert seepweed, which does not have the showy petals typically needed to attract pollinators.

Desert seepweed occurs in California and Arizona east to Texas and as far north as Washington and Canada. In the Grand Canyon, it occurs the length of the river and is especially abundant at Lees Ferry, the delta of Nankoweap Creek, and Palisades Creek.

Of more than one hundred *Suaeda* species found worldwide, two are common in the Grand Canyon, desert seepweed and **woody seepweed (*Suaeda suffrutescens*)**. The main difference is subtle: the leaves of woody seepweed are not strongly flattened and are hairier than those of desert seepweed.

Seepweeds, like sugar beets, spinach, chard, and quinoa, are important food plants in the goosefoot family. Native people gather and grind the small seeds to make *pinole*, a mixture of flour ground from corn, grass, and other seeds. From the mixture they create nutritious cakes that are easily carried while traveling, a kind of ancient granola bar. They eat young, salty greens as an early season food. They also extract a black dye from desert seepweed to decorate pottery and make a poultice from the dried leaves for healing sores.

Tumbleweed *Salsola tragus* —Photograph © 2006 by David Edwards

Desert Seepweed *Suaeda moquinii*
—Photograph © 2006 by Glenn Rink

Desert Seepweed *Suaeda moquinii*
—Photograph © 2006 by Kristin Huisinga

Arizona Grape, Canyon Grape *Vitis arizonica*
GRAPE FAMILY (Vitaceae)

Plants: Creeping, perennial vine, to 6 meters long. **Leaves:** Alternate, broadly cordate, coarsely toothed, to 14 centimeters wide. **Flowers:** In clusters, greenish, inconspicuous. **Fruit:** Clusters of bluish purple berries. **Flowering Season:** April to July. **Elevation:** 2,000 to 7,500 feet.

The woody vines of Arizona grape sprawl and scramble up tree trunks and along canyon and creek bottoms, primarily where there is adequate moisture. Their tendrils are modified leaves that stabilize the plants as they attach to trees and other vegetation, sometimes shading their hosts to death with their verdant foliage.

Arizona grape occurs only in Nevada, Utah, Arizona, New Mexico, and Texas. In the Grand Canyon, it grows along streams and seeps from Nankoweap Creek to Grand Wash Cliffs.

Arizona grape is related to another berry producing, sprawling vine, **Virginia creeper (*Parthenocissus vitacea*)**. Thicket creeper's leaves are divided into five to seven sections, while Arizona grape has undivided leaves with coarse teeth. Less common than Arizona grape in the Grand Canyon, Virginia creeper has a much wider distribution throughout the United States. You may see its foliage turning red in the fall at scattered locations along the river.

The juicy fruits of Arizona grape are used to make jelly, wine, dried fruit, and juice. Dolmas, Middle Eastern staple foods, are rice mixtures wrapped in grape leaves.

Bigelow Onion *Allium bigelovii*
LILY FAMILY (Liliaceae)

Plants: Bulbous perennial herb. **Leaves:** Basal, linear, flattened, extending past flowering stem. **Flowers:** In terminal clusters that are 8 to 12 centimeters tall; six petal-like, whitish pink sepals, with a pink midrib. **Fruit:** Capsule. **Flowering Season:** April to May. **Elevation:** 1,500 to 5,000 feet.

Bigelow onion hails the advent of spring in the Grand Canyon and lucky passersby will see its delicate whitish pink blossoms. Before it blooms, the thin, linear leaves mark the presence of a small, edible onionlike bulb.

Bigelow onion occurs in New Mexico and Arizona. In the Grand Canyon, it grows in open dry soil in the Inner Canyon, especially on the Tonto Plateau, although it has not been well documented in the canyon by botanists.

Bigelow onion resembles another plant from the lily family, **bluedicks (*Dichelostemma pulchellum* var. *pauciflorum*)**, which shares similar habitat and flower shape. Bluedicks flowers are bluish violet, but when in doubt, sniff the leaves for the strong onion scent of Bigelow onion.

The genus name, *Allium*, is Latin for "garlic," and *bigelovii* credits John Milton Bigelow (1804–1878), a botanist and surgeon who traveled with the Pacific Railroad survey in 1853.

Arizona Grape *Vitis arizonica*
—Photograph © 2006 by Bronze Black

Bluedicks *Dichelostemma pulchellum* var. *pauciflorum* —Photograph © 2006 by Max Licher

Virginia Creeper *Parthenocissus vitacea* —Photograph © 2006 by Kristin Huisinga

Bigelow Onion *Allium bigelovii* —Photograph © 2006 by Kristin Huisinga

Mariposa Lily
Calochortus flexuosus
LILY FAMILY (Liliaceae)

Plants: Perennial herb, with corms, to 50 centimeters tall. **Leaves:** Arising from the base, linear, to 20 centimeters long. **Flowers:** Solitary on long stalks, whitish pink to pink, to 4.5 centimeters long. **Fruit:** Three-angled capsule. **Flowering Season:** April to May. **Elevation:** 1,600 to 8,500 feet.

In the spring of the wettest years, mariposa lilies grace Colorado Plateau slopes with hundreds of scattered soft pink blossoms. Three delicate, whitish pink petals form a cup bearing bold-colored nectaries at its center. When not in flower, you might miss these slender, sinuous plants, because only their lilylike leaves are present. Small, edible corms deep beneath the soil surface nourish the developing plant.

Mariposa lily occurs on dry, rocky sites of the desert Southwest from California east to Colorado and New Mexico. In the Grand Canyon, it emerges early in the spring on open desert soils the entire length of the river and to the rims. This exquisite flower resembles few others found at lower elevations, but a close relative, the **white-flowered sego lily (*Calochortus nuttallii*)** is more common at higher elevations.

In earlier times, native people harvested the bulblike corms. They were much needed at the end of harsh winters when few game and food reserves remained. Taking only the largest corms, people replanted the smaller ones in freshly loosened soil. This harvesting technique likely helped maintain or increase the plants' abundance. Mariposa lily buds, flowers, and corms are eaten raw, dried, boiled, steamed and roasted. The genus name, *Calochortus*, means "beautiful grass," referring to its linear leaves that resemble blades of grass.

Desert Bedstraw
Galium stellatum var. *eremicum*
MADDER FAMILY (Rubiaceae)

Plants: Perennial shrub, to 70 centimeters tall. **Leaves:** In whorls of four to five, lanceolate, dark green, to 1 centimeter long. **Flowers:** Cream to greenish yellow, small. **Fruit:** Nutlet with barbed hairs. **Flowering Season:** January to July. **Elevation:** 1,200 to 5,000 feet.

For most of the summer desert bedstraw is dormant and can be difficult to identify because it appears nearly dead. Even then, its needle-tipped leaves arranged in starlike whorls on square stems will help you to differentiate it from other plants. Look for desert bedstraw's characteristically fuzzy, grayish fruits as well. Like many other desert shrubs, male and female flowers occur on separate plants.

Desert bedstraw grows in warm desertscrub communities in California, Nevada, Utah, and Arizona, often on limestone or dolomite outcrops. In the Grand Canyon, it grows directly from cracks in bedrock and on slopes and side canyons the length of the river.

There are nine other species of *Galium* in the Grand Canyon. During a wet spring, you may notice the annual, **sticky willy (*G. aparine*)**, which is a more delicate, trailing species with longer leaves in whorls of six to eight. It grows in shaded, moist habitats from river to rim.

The genus name, *Galium*, is from the Greek word *gala*, meaning "milk," which may refer to the use of another plant in the genus to curdle milk in the manufacture of cheese. The specific epithet is derived from the Latin word *stella*, meaning "starry," referring to the starlike hairs that make this plant sticky to the touch. Legend has it that the manger of baby Jesus was filled with these plants, thus the common name *bedstraw*.

Mariposa Lily *Calochortus flexuosus*
—Photograph © 2006 by David Edwards

Desert Bedstraw *Galium stellatum* var. *eremicum*
—Photograph © 2006 by Kate Watters

Mariposa Lily *Calochortus flexuosus*
—Photograph © 2006 by Regan Dale

Desert Bedstraw *Galium stellatum* var. *eremicum*
—Illustration © 2006 by Mar-Elise Hill

Gooseberryleaf Globemallow *Sphaeralcea grossulariifolia*
MALLOW FAMILY (Malvaceae)

Plants: Perennial herb, to 1.2 meters tall. **Leaves:** Alternate, broadly ovate, coarsely toothed to deeply lobed, prominent veins below, to 5 centimeters wide. **Flowers:** In terminal racemes, reddish orange, 5 to 20 millimeters long. **Fruit:** Wheel-shaped capsule. **Flowering Season:** April to October. **Elevation:** 1,200 to 7,000 feet.

At the end of a wet winter, vast displays of globemallow transform the desert into shades of melon and apricot. Globemallows have five-petaled, reddish orange flowers with stamens joined into a column surrounding the female parts of the flower, an arrangement characteristic of the mallow family.

Globemallows are widespread throughout the western United States. Gooseberryleaf globemallow occurs from Idaho and Washington south to California, Arizona, and New Mexico. In the Grand Canyon, it grows on talus slopes, washes, and in side canyons the length of the river.

There are at least ten different species of globemallow in the Grand Canyon and identifying plants to species can be difficult. **Desert globemallow (*Sphaeralcea ambigua*)**, another common species found near the river, has a more open inflorescence and its leaves extend along the flowering stalk. Gooseberryleaf globemallow has more deeply divided leaves. The distribution of these two species overlaps the length of the river and they likely hybridize, making it even more of a challenge to distinguish between them.

Globemallow seeds are edible and the roots soothe inflamed tissues and promote healing. Native people shred the roots and place them in cold water until they are sticky, then drink the mixture to relieve stomach problems and diarrhea. Globemallow leaves and flower tops, dried and taken as a tea, relieve irritability and promote calmness. The starlike hairs on the foliage can irritate the skin and eyes, hence another common name for this plant, *sore-eye poppy*.

Slender Janusia *Janusia gracilis*
MALPIGHIA FAMILY (Malpighiaceae)

Plants: Perennial vine, to 3 meters long. **Leaves:** Opposite, linear, to 3 centimeters long. **Flowers:** Solitary or in axillary clusters, yellow. **Fruit:** Three-winged, reddish pod. **Flowering Season:** April to October. **Elevation:** 2,300 to 3,500 feet.

Slender janusia twists and crawls over other plants, sometimes forming a rather dense, shrublike cover. Each flower has five yellow petals that look like tiny soup spoons. Even without the flowers, you can distinguish this plant from the few other viny plants found in the canyon by its three-winged, reddish seed pod, lack of milky sap, and hairy, linear leaves.

The malpighia family is mostly a tropical and subtropical family, but slender janusia occurs in Texas, New Mexico, and Arizona, primarily in Sonoran Desert communities. In the Grand Canyon, it grows along Clear and Bright Angel Creeks and at Cremation (River Mile 86), Schist (River Mile 96), and Bass Camps (River Mile 108).

Gooseberryleaf Globemallow
Sphaeralcea grossulariifolia
—Photograph © 2006 by Kate Watters

Gooseberryleaf Globemallow
Sphaeralcea grossulariifolia
—Photograph © 2006 by David Edwards

Slender Janusia *Janusia gracilis* —Photograph © 2006 by Glenn Rink

Broadleaf Milkweed *Asclepias latifolia*
MILKWEED FAMILY (Asclepiadaceae)

Plants: Perennial herb, 20 to 60 centimeters tall. **Leaves:** Opposite, broadly ovate, with a prominent midvein, 4 to 16 centimeters long. **Flowers:** In axillary clusters, white to yellow to pale green. **Fruit:** Podlike follicle. **Flowering Season:** June to August. **Elevation:** 2,700 to 6,400 feet.

The milky sap and distinct clusters of starchy rocket-shaped flowers make all milkweeds easy to recognize. Flowers in the milkweed family have a prominent, central pedestal bearing five hoods that curve over the petals, all perched atop sepals that bend backward. The large, oval-shaped leaves of broadleaf milkweed distinguish it from another, higher-elevation canyon resident, **western whorled milkweed (***Asclepias subverticillata***)**. This plant is thinner-statured with long, narrow, lance-shaped leaves.

The range of broadleaf milkweed includes much of the western United States from California east to Texas and north to South Dakota. In the Grand Canyon, it adorns plains and mesas, especially from upper Marble Canyon to the Hermit Creek area.

Milkweeds are an important food source for monarch butterflies (*Danaus plexippus*) as they pass through the Grand Canyon river corridor on their southward migrations during the late summer and fall. The larvae feed on plants in the milkweed family, accumulating toxins that may help to deter predators. Plants in this genus are known for their purgative, antibiotic, and heart-strengthening properties, but some species have toxic resins that can cause severe poisoning and contact dermatitis. The genus name, *Asclepias*, honors the Greek god of healing, *Asklepios*, who was a prototype of the modern physician.

Twining Milkweed *Sarcostemma cynanchoides* ssp. *cynanchoides*
MILKWEED FAMILY (Asclepiadaceae) (*Funastrum cynanchoides* ssp. *cynanchoides*)

Plants: Twining perennial herb, to 2 meters long. **Leaves:** Opposite, lanceolate to ovate, grayish green to green, lobed at base, to 6 centimeters long. **Flowers:** In umbel-like cymes, white to pink, to 1 centimeter long. **Fruit:** Podlike follicle. **Flowering Season:** April to August. **Elevation:** 1,200 to 3,500 feet.

Twining milkweed trails on almost any plant or object that will support it, such as a metal sign at Lees Ferry or a tree at the water's edge. At first glance, these vines make familiar plants look as if they have evolved odd, new flower forms. The waxy pink flower clusters dangle amidst the foliage of the other plant. The mature seedpods burst open, unleashing the silky threads of their flattened seeds, which cling to neighboring vegetation. When broken, stems and leaves bleed milky sap, an attribute of all plants in the milkweed family.

Twining milkweed trails its way through the deserts of California and Arizona. In the Grand Canyon, it climbs on shrubs and trees the length of the river.

***Sarcostemma hirtellum*,** a less common species of twining milkweed, has greenish yellow flowers and densely hairy leaves that are tapered to a point at both ends. Twining snapdragon (*Maurandella antirrhiniflora*), a smaller vine, has bright green, widely triangular leaves and altogether different flowers, making it easy to discriminate from the twining milkweeds.

The genus name, *Sarcostemma*, is from the Greek words *sarx*, meaning "flesh" and *stemma*, meaning "crown," descriptive of the fleshy, crownlike structure between the petals and the stamens on the flowers of these species.

Broadleaf Milkweed *Asclepias latifolia*
—Photograph © 2006 by David Edwards

Western Whorled Milkweed *Asclepias subverticillata*
—Photograph © 2006 by Kristin Huisinga

Twining Milkweed *Sarcostemma cynanchoides* ssp. *cynanchoides* —Photograph © 2006 by Geoff Gourley

Dwarf Mock Pennyroyal *Hedeoma nanum*
MINT FAMILY (Lamiaceae/Labiatae)

Plants: Annual or perennial herb, to 25 centimeters tall. **Leaves:** Opposite, ovate to elliptic, pale green, red-tinged beneath, with short hairs, 3 to 10 millimeters long. **Flowers:** Whorled in axils, purple, tiny. **Fruit:** Nutlets. **Flowering Season:** February to September. **Elevation:** 1,400 to 6,200 feet.

Dwarf mock pennyroyal is often tucked in the scant soil between rock crevices. Its square stems, opposite leaves, and distinct aroma are characteristic of members of the mint family. The tiny, purple-striped flowers protrude from sepal tubes that linger in the leaf axils long after the petals have fallen. Dwarf mock pennyroyal occurs from Texas west to California and into northern Mexico. In the Grand Canyon, it inhabits dry side canyons and steep rocky slopes the length of the river. Two other species of mock pennyroyal (*Hedeoma drummondii* and *H. oblongifolium*) grow in the Grand Canyon, and all three overlap in range.

Field mint (*Mentha arvensis*), another perennial mint, has square stems and opposite leaves that emit a strong aroma when crushed. The white to lavender flowers are whorled in axils. It grows to 80 centimeters tall with 2- to 8-centimeter-long, serrated leaves. Field mint is widespread throughout the United States. In the Grand Canyon, it thrives in moist areas, often tucked beneath larger plants at Lees Ferry, Buck Farm Canyon, Bright Angel Creek, Tapeats Creek, and Thunder River.

The leaves of dwarf mock pennyroyal are used to soothe stomachaches and colds. The genus name, *Hedeoma*, is from the Greek words *hedus*, meaning "sweet," and *osme*, meaning "odor."

Purple Sage *Salvia dorrii* ssp. *dorrii*
MINT FAMILY (Lamiaceae/Labiatae)

Plants: Shrub, to 80 centimeters tall. **Leaves:** Opposite, linear to spatulate, grayish green, 10 to 35 millimeters long. **Flowers:** Whorled in axils, blue to purple, 6 to 13 millimeters long. **Fruit:** Nutlets. **Flowering Season:** March to May. **Elevation:** 2,000 to 5,500 feet.

The intoxicating aroma of purple sage identifies it as a *true sage*, a member of the *Salvia* genus. The leafy bracts buttress the flower clusters and, along with the flowers, are brilliant blue to pink to purple, fading after pollination. Purple sage flowers have evolved to maximize pollen transfer. The flower's enlarged lower lip provides a landing pad for pollinators, especially bees. Pollen dusts the back of the bee as it gathers nectar. When the bee lands on another flower, the pollen sticks to the stigma, which is perched atop the prominent style.

Purple sage is a regional subspecies of a group called the "*Salvia dorrii* complex," that range from western Canada south to Arizona and Baja California. In the Grand Canyon, purple sage grows on flats and in side canyons such as Cardenas, Stone, and Tapeats.

Another perennial, **Davidson's sage (*Salvia davidsonii*),** is more slender and upright and less woody than purple sage. The pinkish purple flowers are borne on long stalks that rise high above its green, heart-shaped leaves. Its range is restricted to New Mexico and Arizona, and in the Grand Canyon, Davidson's sage is known only from a few side canyons such as Matkatamiba Canyon.

Indigenous people value purple sage for cooking spices, medicines, and cleansing teas. The genus name, *Salvia*, is from the Latin word *salvus*, meaning "healthy" or "safe," which refers to its longstanding use as a medicine.

Dwarf Mock Pennyroyal *Hedeoma nanum*
—Photograph © 2006 by Joe Pollock

Field Mint *Mentha arvensis*
—Photograph © 2006 by Kristin Huisinga

Purple Sage *Salvia dorrii* ssp. *dorrii*
—Photograph © 2006 by Kate Watters

Davidson's Sage *Salvia davidsonii*
—Photograph © 2006 by Raechel Running RMRfotoarts.com

Sahara Mustard
Brassica tournefortii

MUSTARD FAMILY (Brassicaceae/Cruciferae)

Plants: Exotic annual or biennial herb, to 1.5 meters tall. **Leaves:** Basal and cauline, variable, deeply lobed, serrate, hairy, to 50 centimeters long. **Flowers:** In terminal racemes, yellow. **Fruit:** Silique with beaked tip. **Flowering Season:** February to April. **Elevation:** 1,200 to 3,200 feet.

Sahara mustards sprout earlier than most native plants, growing into giants in early spring before other seeds have even germinated. The enormous basal leaves smother surrounding plants and rob them of early spring moisture. As Sahara mustard dies back, its large, dry skeleton rolls across the landscape, spreading sticky seeds far and wide.

Sahara mustard is native to the Mediterranean region, thriving in the broad desert belt from northwestern Africa to the Saudi Arabian peninsula. In North America, it grows from California east to Texas. During the moist spring of 2005, it spread exponentially in the Southwest, earning the nickname "tsunami mustard." Sahara mustard occurs in great numbers in Glen Canyon and Lake Mead National Recreation Areas, but has not yet overwhelmed the Grand Canyon. In the Grand Canyon, it grows on sandy and gravelly soils at scattered locations from Lees Ferry downriver. Park biologists and volunteers began an aggressive control program in the spring of 2004, with a primary focus on the Lees Ferry area.

Two other exotic mustards occur in the Grand Canyon and are also expanding their range. **London rocket (*Sisymbrium irio*)** is shorter than Sahara mustard, with more slender, curved, upright seedpods. **Tumble mustard (*S. altissimum*)** grows as tall as Sahara mustard but its siliques are longer and lack a beaked tip.

Yellow Tansy Mustard
Descurainia pinnata

MUSTARD FAMILY (Brassicaceae/Cruciferae)

Plants: Winter annual herb, to 1 meter tall. **Leaves:** Alternate, bipinnately divided, to 10 centimeters long. **Flowers:** In terminal racemes, yellow. **Fruit:** Silique. **Flowering Season:** March to August. **Elevation:** 1,200 to 7,200 feet.

In the early spring, whole slopes are ablaze with blooming yellow tansy mustards. As other spring plants begin to flower, this plant has already gone to seed, after which its lanky dried stems remain. The plant's form is highly variable, with either a single stem or many branches, depending on the habitat.

Yellow tansy mustard is widespread throughout North America, residing in every state except Alabama. In the Grand Canyon, it is common on both rims, in side canyons, and on slopes and sandy beaches the length of the river. Yellow tansy mustard thrives in recently disturbed, open areas.

At least six other *Descurainia* species grow in the Grand Canyon. **Flixweed (*D. sophia*)** is an exotic annual from Europe and is also common in the Inner Canyon, extending to both rims. Similar in appearance, flixweed is more often branched, has tripinnate lower leaves and erect siliques that are nearly twice as long as those of yellow tansy mustard.

Yellow tansy mustard provides food for several butterfly species in their larval stage. Mammals consume the young leaves, although it can be toxic in large quantities. Yellow tansy mustard has a long history of use as an edible spring green and the seeds can be used as a spice. The genus name, *Descurainia*, honors the French botanist Francois Descourain (1658–1740).

Sahara Mustard *Brassica tournefortii*
—Photograph © 2006 by Max Licher

Sahara Mustard *Brassica tournefortii*
—Photograph © 2006 by Max Licher

London Rocket *Sisymbrium irio*
—Photograph © 2006 by Kate Watters

Yellow Tansy Mustard *Descurainia pinnata*
—Photograph © 2006 by Glenn Rink

Wedge-Leaf Draba *Draba cuneifolia*
MUSTARD FAMILY (Brassicaceae/Cruciferae)

Plants: Annual herb, to 40 centimeters tall. **Leaves:** Mostly basal, oblanceolate to obovate, coarsely toothed, hairy, to 7 centimeters long. **Flowers:** In racemes, white, tiny. **Fruit:** Silique. **Flowering Season:** February to May. **Elevation:** 1,200 to 7,000 feet.

Wedge-leaf draba grows in abundance in the late winter and early spring, often blanketing the ground. While most often seen in open areas, these tiny plants sometimes tuck themselves beneath the canopy of trees and shrubs. Look for its small white flowers growing in dense clusters poised atop the densely hairy stems and leaves.

Wedge-leaf draba is widespread in the United States, occurring in almost every southern state. In the Grand Canyon, it grows the length of the river and extends to the South Rim.

The genus name, *Draba*, is from the Greek word *drape*, meaning "acrid," referring to the bitter taste of the leaves. The specific epithet, *cuneifolia*, refers to the wedge-shaped leaves.

Desert Pepperweed *Lepidium fremontii*
MUSTARD FAMILY (Brassicaceae/Cruciferae)

Plants: Perennial herb, to 1 meter tall. **Leaves:** Alternate, linear, entire to lobed, to 10 centimeters long. **Flowers:** In panicles, white. **Fruit:** Silique. **Flowering Season:** March to September. **Elevation:** 1,500 to 3,700 feet.

Robust, shrubby desert pepperweeds cover the upper beaches and slopes of Marble Canyon during spring. These showy plants are cloaked in thousands of tiny white blossoms, the mass of white visible even from the top of the trail to Saddle Canyon. Typical of the mustard family, desert pepperweed has four-petaled flowers with six stamens.

Desert pepperweed grows in Utah and Arizona. It inhabits disturbed areas and openings in woodlands and forests. In the Inner Canyon, if you see a pepperweed, it is likely desert pepperweed, which occurs the entire length of the river.

Differentiating among pepperweeds can be tough, especially without fruit, because they have similar forms and overlapping ranges. In the Inner Canyon, **mountain pepperweed (*Lepidium montanum*)** is less common than desert pepperweed, but at higher elevations, it becomes the dominant of the two species. Its stems are less woody above the base, with leaves to 12.5 centimeters long, narrower flowering stalks and larger fruits. The exotic **broad-leaved pepperweed (*L. latifolium*)** was first collected in Grand Canyon National Park in 1993 and is spreading rapidly into moist soils along the river corridor. Its broad, entire leaves have toothed margins, unlike the other two species, and its flowering stalks are more open and showy.

Known as the poor man's pepper, you can use the dried pods of pepperweeds as a substitute for black pepper and the greens will add spice to any salad.

Wedge-Leaf Draba *Draba cuneifolia*
—Photograph © 2006 by David Edwards

Broadleaved Pepperweed *Lepidium latifolium*
—Photograph by Lori Makarick/National Park Service

Broadleaved Pepperweed *Lepidium latifolium* —Photograph by Lori Makarick/National Park Service

Desert Pepperweed *Lepidium fremontii* —Photograph © 2006 by David Edwards
Inset —Photograph © 2006 by Geoff Gourley

Watercress
Rorippa nasturtium-aquaticum
MUSTARD FAMILY (Brassicaceae/Cruciferae)   (*Nasturtium officinale*)

Plants: Exotic, semiaquatic perennial herb, with succulent stems, to 1 meter long. **Leaves:** Alternate, ovate, simple to pinnately divided, dark green, to 10 centimeters long. **Flowers:** In racemes, white. **Fruit:** Silique. **Flowering Season:** March to August. **Elevation:** 2,000 to 5,200 feet.

The submersed stems and roots of watercress grow horizontally, float, and sprout freely. You have probably seen watercress in your own local ponds and streams, growing just as it grows in the Grand Canyon. It provides valuable shelter and food for the endangered Kanab ambersnail (*Oxyloma haydeni kanabensis*) at Vasey's Paradise.

Watercress, native to Europe, has become very widespread in North America. In the Grand Canyon, it frequents seeps, springs, and moist sites between Vasey's Paradise and Havasu Canyon.

The peppery-tasting leaves of watercress are high in vitamins A and C and are often used in salads and soups. Care should be taken, however, because eating leaves from bacteria-laden waters can cause illness.

Prince's-Plume
Stanleya pinnata var. *pinnata*
MUSTARD FAMILY (Brassicaceae/Cruciferae)

Plants: Perennial herb, to 1.5 meters tall. **Leaves:** Alternate, lanceolate to pinnately lobed, blue green, 5 to 20 centimeters long. **Flowers:** In racemes, yellow. **Fruit:** Dangling silique. **Flowering Season:** May to July. **Elevation:** 2,200 to 7,600 feet.

The clustered, bright yellow flowers of prince's-plume are conspicuous at the top of long, graceful flowering stalks that sway in the breeze and hold the glow of the setting sun. Prince's-plume has a white-flowered look-alike, **thelypody (*Thelypodium integrifolium*)**, with similarly arranged flowers, but thelypody has thick, entire leaves and a taller, more slender stature, towering above the understory.

Four varieties of prince's-plume occur throughout the majority of the western United States. In the Grand Canyon, prince's-plume grows on soils derived from shales, mudstones, and siltstones throughout the Inner Canyon.

Prince's-plume grows in soils rich in selenium, a naturally occurring element that is hazardous at high levels. Without damaging its tissues, the plants accumulate selenium and convert it into a nonhazardous gas, releasing it into the atmosphere. By feeding on the plants, insects transfer selenium from the plant tissue to organisms higher in the food chain where it could accumulate to toxic levels. Although the greens are considered poisonous, native people eat them in small amounts, boiling them two or three times to remove the bitter taste. A poultice of mashed roots treats rheumatic pains, earaches, and toothaches. The genus name, *Stanleya*, commemorates Lord Edward Smith Stanley (1773–1849), an ornithologist and past president of the Linnaean Society.

Watercress *Rorippa nasturtium-aquaticum* —Photograph © 2006 by Kate Thompson
Inset —Photograph © 2006 by Kate Watters

Prince's-Plume *Stanleya pinnata* var. *pinnata*
—Photograph © 2006 by Kristin Huisinga

Thelypody *Thelypodium integrifolium* ssp. *longicarpum*
—Photograph © 2006 by Kristin Huisinga
Inset —Photograph © 2006 by Max Licher

Long-Beaked Twist Flower   *Streptanthella longirostris*
MUSTARD FAMILY (Brassicaceae/Cruciferae)

Plants: Annual herb, to 50 centimeters tall. **Leaves:** Alternate, narrowly oblanceolate, entire to toothed, to 8.5 centimeters long. **Flowers:** In racemes, white to yellow, with purple veins. **Fruit:** Silique. **Flowering Season:** January to June. **Elevation:** 1,200 to 8,600 feet.

When seeds of other annual plants are still awaiting ideal conditions, long-beaked twist flower seeds have already germinated. Taking advantage of winter rains, this plant grows abundantly in the early spring, often living amidst biological soil crusts and in open, sandy soils. The tall swaying stems display distinctive, urn-shaped flowers streaked with purple veins.

Long-beaked twist flower occurs throughout the western United States. In the Grand Canyon, it grows the length of the river on beaches and rocky slopes. It also flourishes at high elevations beneath dense forest canopies.

The genus name, *Streptanthella*, is from the Greek words *streptos*, meaning "twisted," *anthus*, meaning "flower," and *ella*, for its diminutive nature. The specific epithet, *longirostris*, is Latin for "long beak," referring to the silique.

Sacred Datura, Jimson Weed   *Datura wrightii*
NIGHTSHADE FAMILY (Solanaceae)   (*Datura meteloides*)

Plants: Perennial herb, to 1.5 meters tall. **Leaves:** Alternate, triangular to broadly ovate, 4 to 15 centimeters long. **Flowers:** Solitary, tubular, white with hints of purple. **Fruit:** Prickly, round capsule. **Flowering Season:** May to October. **Elevation:** 1,200 to 4,200 feet.

Sacred datura graces beaches and slopes with its enormous, moon-colored flowers that open in the evening, attracting a variety of night-flying pollinators. The funnel-shaped flowers shrivel and droop in the heat of the day. Even without the conspicuous flowers, the dark green leaves and large stature of this spreading plant distinguishes it from other desert plants. The leaves contain high concentrations of toxic alkaloids that can cause blindness, kidney failure, and death, so admire sacred datura's beauty from afar.

Members of the genus occur worldwide and are widespread throughout North America. Our species is restricted to the Southwest. In the Grand Canyon, sacred datura blooms prolifically all year long on slopes and in side canyons throughout the river corridor.

Creatures that look like a cross between a hummingbird and a moth, called sphinx or hawk moths (Sphingidae), are particularly fond of these intoxicating flowers. Hawk moths and sacred datura exhibit an intricate codependence. Hawk moth larvae—large, green hornworms—feast on the alkaloid-laden leaves of sacred datura. By ingesting these toxins, the juicy hornworms gain protection from predators who quickly learn to avoid them. In return, adult hawk moths are one of the primary pollinators for sacred datura, providing the transfer of pollen. Few other pollinators have a long enough proboscis to reach the nectar that lies deep within the base of the flower tube. Look for dizzy hawk moths, drunk on sacred datura's alkaloids, circling about in the early evening.

The common name, Jimson weed, shortened from Jamestown weed, comes from a 1676 incident in Jamestown, Virginia. British soldiers were sent to quell an uprising and upon their arrival, they were sick with scurvy and fed a mixture of leafy greens. Among the mixture was Jimson weed (*Datura stramonium*), which temporarily debilitated them. Despite its toxic qualities, medicine people throughout the world employ this plant during ceremonial rites.

Long-Beaked Twist Flower *Streptanthella longirostris* —Photograph © 2006 by Geoff Gourley

Flowers and developing fruit of Long-Beaked Twist Flower *Streptanthella longirostris* —Photograph © 2006 by Glenn Rink

Sacred Datura *Datura wrightii* —Photograph © 2006 by Bob Rink

Green hornworm feeding on Sacred Datura *Datura wrightii* —Photograph © 2006 by Charly Heavenrich

WOLFBERRY *Lycium* species
NIGHTSHADE FAMILY (Solanaceae)

Anderson's Wolfberry *Lycium andersonii*
Plants: Rounded shrub, to 3 meters tall. **Leaves:** Alternate or clustered, linear, fleshy, 3 to 15 millimeters long. **Flowers:** Solitary or in axillary pairs, tubular, white to pale lavender. **Fruit:** Orange to red berry. **Flowering Season:** February to May. **Elevation:** 1,300 to 6,500 feet.

Pale Wolfberry *Lycium pallidum*
Plants: Spreading shrub, to 2 meters tall. **Leaves:** Alternate or clustered, ovate to spatulate, 1 to 4 centimeters long. **Flowers:** Solitary or in axillary clusters, tubular, green with purple tinge. **Fruit:** Bright red or blue berry. **Flowering Season:** April to June. **Elevation:** 2,400 to 6,300 feet.

Like many desert shrubs, wolfberries appear lifeless during the hot, dry summer months when they drop their foliage, leaving only thorny branches. When moisture arrives, the succulent green leaves emerge and in the spring small flowers bloom.

Leaf and fruit size, as well as overall architecture, differentiate these two shrubs. Pale wolfberry has larger, spatulate leaves and more arching, flexible stems with fewer thorns. Anderson's wolfberry has succulent, linear leaves and thornier, stiffer branches. An unrelated species in the rose family, desert almond (*Prunus fasciculata*), can be mistaken for wolfberry when only leaf bundles and spiny branches are present. Look for tiny, white, five-petaled, nontubular flowers and generally darker bark to positively identify desert almond.

Anderson's wolfberry is a common Mojave Desert plant, occurring in California, Nevada, Utah, Arizona, New Mexico, and northern Mexico. In the Grand Canyon, it grows in gravelly washes and on sandy flats from North Canyon downstream. It is the species you are most likely to encounter near the river. Pale wolfberry has a wider distribution in the Southwest. In the Grand Canyon, pale wolfberry is found on both rims, along the Hermit, Hance, and Nankoweap Trails, and in a few side canyons, including Havasu Canyon. Another species, **Fremont wolfberry (*Lycium fremontii*),** is less common, reported only from five localities in the Inner Canyon. While it also has linear, succulent leaves, the glands covering the leaves and flower tube set it apart.

Black-chinned hummingbirds (*Archilochus alexandri*) visit the nectar-rich blossoms. The flowers mature into small red berries that are savored by birds and small mammals. Larger wildlife browse the leaves and stems, encouraging bushy growth. The vitamin-rich berries were a staple food for native people, who mixed them with clay to temper the bitter taste. They are still eaten today. Wolfberry plants are often associated with archaeological sites.

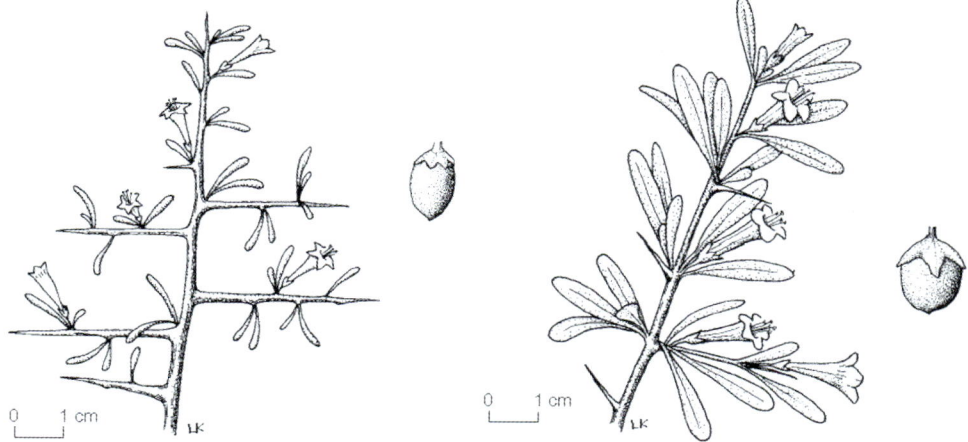

Anderson's Wolfberry *Lycium andersonii*
—Illustration © 2006 by Lisa Kearsley

Pale Wolfberry *Lycium pallidum*
—Illustration © 2006 by Lisa Kearsley

Anderson's Wolfberry
Lycium andersonii
—Photograph © 2006
by Lisa A. Hahn

Pale Wolfberry
Lycium pallidum
—Photograph © 2006
by Kate Watters

Desert Tobacco *Nicotiana obtusifolia* var. *obtusifolia*
NIGHTSHADE FAMILY (Solanaceae) (*N. trigonophylla*)

Plant: Annual or short-lived perennial, to 1.2 meters tall. **Leaves:** Alternate, lanceolate, clasping at base, to 15 centimeters long. **Flowers:** In racemes or panicles, tubular, cream-colored, to 2.5 centimeters long. **Fruit:** Capsule. **Flowering Season:** April to September. **Elevation:** 1,200 to 8,400 feet.

In the spring and early summer, desert tobacco stands out as one of the more vigorous plants in drier areas. The glandular leaves emit a distinctive, stinky aroma, which helps identify this plant in the absence of its creamy vanilla, tubular flowers.

Desert tobacco occurs from California east to Oklahoma and Texas, with a disjunct population in Maryland. In the Grand Canyon, desert tobacco grows among rocks, along washes, and in disturbed areas the length of the river.

For centuries, people have used desert tobacco for just about everything that ails. The leaves can stop bleeding or itching and ease sore muscles, bruises, and sunburn. The leaves are dried and crushed, then smoked or remoistened and molded into plugs that are chewed, swallowed, or snuffed. Native tobaccos are central in many Native American ceremonies. In addition to nicotine, desert tobacco contains the highly toxic alkaloid anabasine, which is used as an insecticide. The genus name, *Nicotiana*, honors Jean Nicot (1530–1600), who reportedly introduced tobacco to sixteenth-century France.

Thick-Leaved Ground Cherry *Physalis crassifolia*
NIGHTSHADE FAMILY (Solanaceae)

Plants: Perennial herb, 20 to 80 centimeters tall. **Leaves:** Alternate, ovate to cordate, 1 to 5 centimeters long. **Flowers:** Solitary or in clusters, bell-shaped, dull yellow. **Fruit:** Berry, surrounded by papery calyx. **Flowering Season:** March to October. **Elevation:** 1,200 to 7,400 feet.

When you first spot thick-leaved ground cherry, you might mistake it for its cultivated relative, tomatillo (*Physalis ixocarpa*). Both species have papery, lanternlike fruits hanging from their branches, but thick-leaved ground cherry's are much smaller. Like the tomatillo, the yellow mature fruits are edible, but unripe fruits are toxic in the wild species.

Thick-leaved ground cherry is found from Colorado and Texas west to California and into Mexico. In the Grand Canyon, it is common the length of the river, primarily in side canyons like those of Stone and Deer Creeks.

Ivy-leaved ground cherry (*P. hederifolia*) is another common *Physalis* species in the Inner Canyon. Although similar in size to thick-leaved ground cherry, it is more sparsely branched and its yellow flowers have dark spots at the base.

Ground cherries provided important food for desert-dwelling Native Americans. Where plentiful, it may indicate the presence of Ancestral Puebloan garden sites.

Desert Tobacco
Nicotiana obtusifolia var. *obtusifolia*
—Photograph © 2006 by Kristin Huisinga

Desert Tobacco *Nicotiana obtusifolia* var. *obtusifolia*
—Photograph © 2006 by Kathy Darrow

Thick-Leaved Ground Cherry *Physalis crassifolia* —Photograph © 2006 by Lisa A. Hahn

Silverleaf Nightshade
Solanum elaeagnifolium

NIGHTSHADE FAMILY (Solanaceae)

Plants: Exotic perennial herb, to 1 meter tall. **Leaves:** Simple, linear to oblong, with wavy margins, to 15 centimeters long. **Flowers:** In panicles, star-shaped, white to violet. **Fruit:** Yellow to black berry. **Flowering Season:** May to October. **Elevation:** 2,400 to 3,800 feet.

Silverleaf nightshade has distinctive silvery leaves, but the nodding, purple flowers with protruding bright yellow anthers are its most conspicuous features. This plant thrives in disturbed areas and grows readily along the trails near Phantom Ranch. Deep underground stems extend far from the parent plant and produce new shoots.

Silverleaf nightshade is native to the central United States and has expanded its range into other regions, including the Grand Canyon. Well adapted to semiarid regions and on sandy soils, it is at home in the Grand Canyon's low-elevation sites.

At least seven other *Solanum* species occur in the Grand Canyon including the native **American nightshade (*S. americanum*),** which is a robust, dark green annual with white flowers and greenish black fruits. Look for it at Deer Creek, Cove Canyon, and other side canyons.

Native people used crushed berries to curdle milk in cheese making and seeds to tan hides. Silverleaf nightshade is listed as a noxious weed in twenty-one states, including Arizona, because it overtakes whole areas and is poisonous to livestock.

Beargrass
Nolina microcarpa

NOLINA FAMILY (Nolinaceae)
(Formerly in the Agave Family)

Plants: Succulent perennial, to 2 meters tall. **Leaves:** In a rosette, linear, minutely toothed margins, stringy at tip, to 1.2 meters long. **Flowers:** In panicles that are up to 1 meter tall; flowers creamy white, to 6 millimeters long. **Fruit:** Dry, winged capsule. **Flowering Season:** April to July. **Elevation:** 1,400 to 6,900 feet.

Beargrass is easily identified by its large clump of grasslike leaves and tall flowering stalks that persist long after the creamy vanilla flowers become fruit. Although beargrass appears to be a large grass, it is not a member of the grass family. From a distance, beargrass also resembles sawgrass (*Cladium californicum*), which is a member of the sedge family and has rusty brown, drooping flower clusters on long stalks. The deep green leaves of sawgrass have coarser teeth and lack the curly, frayed leaf tip of beargrass.

Beargrass occurs from southwestern Utah to northern Mexico and is especially common on rocky hillsides in Arizona and New Mexico. In the Grand Canyon, plants grow on rocky, open slopes, in alcoves, and along waterways. Look for it near Phantom Ranch and downstream to Grand Wash Cliffs. It is particularly abundant on the Esplanade Formation. Grand Canyon plants have wider, deeper green leaves, more robust flowering stalks, and larger fruits than *Nolina microcarpa* growing elsewhere in the Southwest.

The sturdy leaves of beargrass were tradeable commodities used for the fabrication of brushes, cords, rope, whips, and mats. Even today, 90 percent of the beargrass fiber processed in Sonora and Chihuahua, Mexico, is sold to broom supply companies in the United States. The young stalks were eaten as a vegetable, the seeds ground into flour, and the roots used for treating pneumonia or rheumatism.

Silverleaf Nightshade *Solanum elaeagnifolium*
—Photograph © 2006 by Bronze Black

American Nightshade *Solanum americanum*
—Photograph © 2006 by Max Licher

Beargrass *Nolina microcarpa* —Photograph © 2006 by Charly Heavenrich

Ocotillo
Fouquieria splendens

OCOTILLO FAMILY (Fouquieriaceae)

Plants: Thorny, semi-succulent shrub, to 9 meters tall. **Leaves:** In clusters, obovate, 1 to 2 centimeters long. **Flowers:** In terminal panicles, tubular, orangish red. **Fruit:** Capsule. **Flowering Season:** February to August. **Elevation:** 1,200 to 2,300 feet.

With erect, swordlike branches reaching toward the sky, ocotillo is one of the most characteristic plants of the southwestern deserts. This partially succulent shrub comes alive in the spring when brilliant orangish red flower clusters burst from its branch tips and lush, green foliage dresses its stems. After the rains subside, the leaves drop and its gray branches blend into the surrounding landscape.

Ocotillo, with its thorny, often leafless stems, resembles a cactus or a succulent, but it is neither. It has, however, evolved strategies similar to cacti, such as shallow roots, reduced leaves, photosynthetic stems, and water storage organs, to cope with the extreme heat and dryness of harsh desert environments. In an attempt to conserve water during dry spells, ocotillo may shed and grow its foliage five or six times in one year.

Thirteen species of *Fouquieria* are restricted to arid regions in North America. *F. splendens* grows on slopes and mesas in grasslands and deserts from California east to Texas and south into mainland Mexico and Baja California. In the Grand Canyon, ocotillo grows from River Mile 155 downstream to Grand Wash Cliffs. At higher elevations outside the Grand Canyon, it favors limestone substrates, which are able to retain warmth longer than other rocks, thus allowing it to persist in cooler areas at the northern edge of its range.

Hummingbirds in need of fast-food stopovers during spring migration feed on ocotillo, which is often the only plant blooming in drought years. The tubular flowers have coevolved to suit the needs of its specific pollinators. Carpenter bees (*Xylocopa* species) crawl from flower to flower gathering pollen as they feed on the nectar tubes. Antelope ground squirrels (*Ammospermophilus* species) scurry up ocotillo branches and feed on the seeds and flowers.

Throughout its range, native people, pioneers, and explorers have found many uses for ocotillo. They use the branches for house construction, firewood, and torches. Branch cuttings root readily to make living fences. The flowers, soaked in cold water, make a very refreshing, tasty beverage. Herbalists make a tea from the bark to cleanse the lymph system and as a poultice to reduce swelling. The genus name, *Fouquieria*, honors Parisian professor of medicine Pierre Éloi Fouquier (1776–1850), and the specific epithet, *splendens*, is descriptive of the brilliant flowers. Ocotillo has at least sixteen common names, many of which are derived from Spanish, Aztec, and Native American languages.

Ocotillo *Fouquieria splendens*
—Photograph © 2006 by Kate Thompson

Ocotillo *Fouquieria splendens*
—Photograph © 2006 by John Running

Ocotillo *Fouquieria splendens* —Photograph © 2006 by Regan Dale

Roundleaf Buffaloberry
Shepherdia rotundifolia
OLEASTER FAMILY (Elaeagnaceae)

Plants: Evergreen shrub, to 2 meters tall. **Leaves:** Opposite, ovate, gray green, to 4 centimeters long. **Flowers:** In axils, greenish yellow, inconspicuous. **Fruit:** Dry, olive-shaped berry. **Flowering Season:** March to June. **Elevation:** 3,200 to 8,500 feet.

Roundleaf buffaloberry, a striking shrub with silvery gray foliage, stands out against the brilliant cliffs and slopes of the Grand Canyon. The thick leaves with down-curved margins are covered with star-shaped yellow or whitish hairs that feel like felt. The leaves emit a faint and pleasant, lemony fragrance. Separate from the males, the female plants develop dry fruits, unlike the edible but tart fruits of its cousin **silver buffaloberry (*Shepherdia argentea*)**, which grows on the North Rim. Both species share the same family as Russian olive (*Elaeagnus angustifolia*), an invasive tree that dominates many southwestern riparian areas.

Roundleaf buffaloberry grows in desertscrub, piñon-juniper, and ponderosa pine communities, restricted to southern Utah and northern Arizona. In the Grand Canyon, it adorns open rocky slopes in sandstone substrates at scattered locations from the Inner Canyon to the rims. Along the river corridor, plants form compact, silver mounds in the pre-dam high-water zone from Marble Canyon to Havasu Creek.

Havasupai people say the dust, or hairs, from the leaves makes the eyes sore and causes blindness. Navajo people burn the plant's aerial parts and use the ash as a lotion for headaches, to heal navels, and to treat toothaches. Roundleaf buffaloberry is one of only three species within this genus, named for John Shepherd (1764–1836), who was a botanist and curator of the Liverpool Botanic Gardens.

Helleborine Orchid
Epipactis gigantea
ORCHID FAMILY (Orchidaceae)

Plants: Rhizomatous perennial herb, to 1.4 meters tall. **Leaves:** Alternate, lanceolate to ovate, clasping at base, 5 to 20 centimeters long. **Flowers:** In racemes, yellowish green to brownish purple. **Fruit:** Capsule. **Flowering Season:** April to July. **Elevation:** 1,200 to 7,600 feet.

The orchid family is one of the largest and most diverse plant families worldwide, with nearly 20,000 species from 800 genera. The highly evolved intricate structure of orchid flowers is designed to attract pollinators. Despite this, pollination in nature is rare. When pollinated, plants pack millions of tiny, microscopic seeds into one fruit. Even without helleborine orchid's distinctive flowers, you can identify the plant by its lance-shaped, parallel-veined leaves that clasp the stem. Look for these plants clinging to precarious cliff faces and overhangs in the spray of waterfalls, where they seek out moist, cool, shady homes.

Helleborine orchid occurs across most of western North America from southern Canada to northern Mexico. Arizona is home to twenty-six orchid species. One of three orchid species found in the Inner Canyon, helleborine orchid is the most common and widely distributed, extending to moist locations on the rims. It grows in seeps, springs, and hanging gardens along the river and in side canyons from Vasey's Paradise to Grand Wash Cliffs.

Roundleaf Buffaloberry *Shepherdia rotundifolia* —Photograph © 2006 by Tina Ayers

Helleborine Orchid *Epipactis gigantea* —Photograph © 2006 by Roger Dale

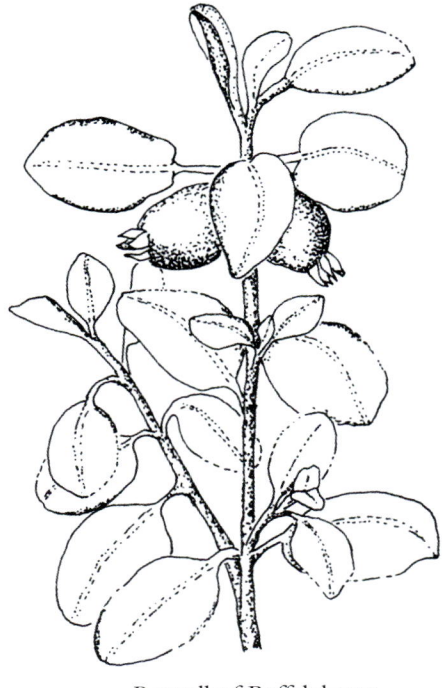

Roundleaf Buffaloberry
Shepherdia rotundifolia
—Illustration © 2006 by Mar-Elise Hill

Camelthorn
Alhagi maurorum
(*Alhagi camelorum*)

PEA FAMILY (Fabaceae/Leguminosae)

Plants: Exotic perennial herb or shrub, to 1 meter tall. **Leaves:** Alternate, elliptic to obovate, 7 to 20 millimeters long. **Flowers:** In racemes, pink to reddish purple, to 1 centimeter long. **Fruit:** Pod. **Flowering Season:** May to July. **Elevation:** 1,200 to 2,800 feet.

Camelthorn, a highly invasive plant, forms dense, spiny thickets along rivers with its extensive rhizome and root system. What you see aboveground is only a small fraction of its true mass, the bulk of which grows deep below the soil surface. Little can be done to control this hardy plant. Pulling individual plants encourages new shoots to sprout from its underground rhizomes. River travelers will undoubtedly curse camelthorn as it pokes and scrapes their legs during their quest for the perfect campsite.

Native to Eurasia, camelthorn was unintentionally brought to the United States in packing material used to transport another exotic plant, date palm (*Phoenix dactylifera*). Camelthorn is spreading rapidly in the southwestern United States and currently grows from California and Nevada east to Texas. It reached the Grand Canyon in the late 1960s via plant fragments that traveled down the Little Colorado River. Even today, it is found only downstream of the Little Colorado River, where it continues to colonize beaches and sandbars with great vigor all the way to Lake Mead.

Camelthorn is disliked in our region for its habit of crowding out native vegetation, but Middle Eastern nomads value and collect it for spring forage. Camels feed on young shoots, and humans may resort to eating its fibrous roots during times of famine.

Specklepod Milkvetch
Astragalus lentiginosus

PEA FAMILY (Fabaceae/Leguminosae)

Plants: Perennial or biennial herb, 10 to 100 centimeters tall. **Leaves:** Pinnately compound, with nine to twenty-seven oblong to ovate leaflets, 1 to 17 centimeters long. **Flowers:** In racemes, pinkish purple, 7 to 22 millimeters long. **Fruit:** Variously inflated pod. **Flowering Season:** February to August. **Elevation:** 1,200 to 7,000 feet.

The genus *Astragalus* is extremely diverse with over 2,500 species and numerous varieties worldwide. If you encounter an herbaceous plant with pinnately compound leaves, pealike flowers, and beanlike pods, it is most likely an *Astragalus*. Identifying individual species is difficult, even with the help of a microscope, because only subtle differences distinguish them.

Specklepod milkvetch occurs from California and Oregon east to Wyoming and south to Mexico. It is the most common perennial *Astragalus* species in the Grand Canyon, occurring from river to rim in a variety of habitats, but preferring sandy soils. More than twenty-one species of *Astragalus* grow in the Grand Canyon, and at least seventeen are found either along the Colorado River or in adjacent side canyons. The Hermit Trail, Clear Creek, and Havasu Canyon are particularly rich in milkvetch species. **Stinking milkvetch (*A. praelongus*)**, a more upright perennial, has dark green leaves and white flowers. **Small-flowered milkvetch (*A. nuttallianus*)** is a prostrate, purple-flowered annual common throughout the Inner Canyon.

Most species in the genus *Astragalus* are milkvetches. Another group, called *locoweeds*, contain toxins that negatively affect the nervous systems of wildlife and livestock. The genus name is from the Greek word *astragalos*, meaning "anklebone." Ancient Greeks used the bones from the ankle as dice, and when shaken, they rattled much like the seeds in an *Astragalus* pod.

Camelthorn *Alhagi maurorum*
—Photograph © 2006 by Glenn Rink

Top: **Stinking Milkvetch** *Astragalus praelongus* —Photograph © 2006 by Glenn Rink
Bottom: **Small-Flowered Milkvetch**
Astragalus nuttallianus —Photograph © 2006 by Max Licher

Specklepod Milkvetch *Astragalus lentiginosus* —Photograph © 2006 by Geoff Gourley

Yellow Sweetclover
Melilotus officinalis

PEA FAMILY (Fabaceae/Leguminosae)

Plants: Exotic annual or biennial herb, to 2 meters tall. **Leaves:** Palmately compound, with three oblong, toothed leaflets, 1 to 2.5 centimeters long. **Flowers:** In spikelike racemes, yellow, to 1 centimeter long. **Fruit:** Pod. **Flowering Season:** May to October. **Elevation:** 1,200 to 8,000 feet.

Yellow sweetclover, a common widespread plant, has three-parted, cloverlike leaves. Like other members of the pea family, its roots have tiny nodules that house *Rhizobium* bacteria, which convert atmospheric nitrogen into a usable form for nearby plants. Sweetclovers (*Melilotus* species) are prolific seed producers with one plant yielding over 250,000 seeds. Though sweetclovers, like many non-native plants, readily establish on disturbed sites, they generally do not persist after other vegetation becomes established.

Sweetclovers were introduced from Eurasia as early as 1664 to be used as forage, nectar crops, and soil builders. Both drought and cold tolerant, they are adapted to a wide range of climatic conditions, and their seeds remain viable in the soil for many years. In the Grand Canyon, yellow sweetclover grows in moist sandy soils along the river and in side canyons throughout the Inner Canyon, extending to the canyon rims. In the same habitats, you may notice a similar plant that has white flowers. This is **white sweetclover (*Melilotus alba*)**. When not in flower, sweetclovers resemble **alfalfa (*Medicago sativa*)**, which has purple flowers.

Sweetclovers are excellent honey plants because of their sweet-smelling, attractive flowers that are pollinated by many different types of bees. Young shoots provide valuable forage. Dried plants emit a faint vanilla-like odor, which is due to the presence of coumarin, a plant chemical used in perfumes and some drugs. The genus name, *Melilotus*, is derived from the Greek roots *meli*, meaning "honey," and *lotos*, meaning "a leguminous plant."

Common Dunebroom
Parryella filifolia

PEA FAMILY (Fabaceae/Leguminosae)

Plants: Shrub, to 1.5 meters tall. **Leaves:** Alternate, pinnately compound, with eleven to forty-five linear leaflets, to 13 centimeters long. **Flowers:** In racemes, yellow, tiny. **Fruit:** Pod. **Flowering Season:** June to September. **Elevation:** 2,600 to 6,500 feet.

The genus *Parryella* contains only one species, common dunebroom. All parts of this shrub, especially the pods, are dotted with glands and smell like citrus when crushed. These shrubs often appear only as a collection of green stems because they shed their leaves in response to drought.

Common dunebroom occurs only on the southern Colorado Plateau in Colorado, Utah, Arizona, and New Mexico. Despite its name, common dunebroom is a rare find in the Grand Canyon, growing only at Lees Ferry and scattered locations downstream. It is visible at Trinity Canyon.

Hopi people use common dunebroom for insecticide, cordage, and as one of the materials for wicker baskets.

Yellow Sweetclover *Melilotus officinalis* —Photograph © 2006 by Kristin Huisinga

Alfalfa *Medicago sativa*
—Photograph © 2006 by Glenn Rink

Common Dunebroom *Parryella filifolia*
—Illustration © 2006 by Mar-Elise Hill

Rushlike Scurf-Pea — *Psoralidium junceum*
PEA FAMILY (Fabaceae/Leguminosae)   (*Psoralea juncea*)

Plants: Perennial herb, to 1 meter tall. **Leaves:** Alternate, palmately compound, reduced. **Flowers:** In clusters, dark blue to purple, to 6.5 millimeters long. **Fruit:** Circular, silky hairy pods. **Flowering Season:** May. **Elevation:** 1,950 to 2,350 feet.

Rushlike scurf-pea forms clumps of many bright green, nearly leafless stems topped by delicate, indigo, pea-shaped blossoms. It is endemic to the Colorado and San Juan River drainages and many populations were likely drowned by the rising waters of Lake Powell, which inundated habitat for this plant. Today, it is only known from a few locations below Glen Canyon Dam where it grows in deep, sandy soils.

The genus name, *Psoralidium*, is a diminutive of *Psoralea*, which means "roughly scaled" in Greek, referring to the dotlike glands on the leaves and stems of plants in this genus. The specific epithet, *junceum*, translates to "rushlike," in reference to their resemblance to rushes, which also grow in clumps with round green stems.

Mojave Indigobush — *Psorothamnus arborescens*
PEA FAMILY (Fabaceae/Leguminosae)   (*Dalea arborescens*)

Plants: Shrub, to 1 meter tall. **Leaves:** Alternate, pinnately compound, with five to fifteen leaflets, to 4 centimeters long. **Flowers:** In racemes, violet purple, to 1 centimeter long. **Fruit:** Gland-dotted pod. **Flowering Season:** April to May. **Elevation:** 1,200 to 4,400 feet.

In early spring, Mojave indigobush decorates the rocky desert slopes with its rich, saturated purple blossoms and silvery green foliage. For most of the year, this stout, intricately branched shrub is dormant and leafless. You can still identify it in its dormant state because glands dot the stems and emit a pungent odor.

Mojave indigobush lives among other desert shrubs in California, Nevada, and Arizona, and south to Mexico. In the Grand Canyon, it occurs from South Canyon to Grand Wash Cliffs.

Fremont indigobush (*Psorothamnus fremontii*), another species with an overlapping range, is very similar, but grows to 1.5 meters tall and has a slightly different calyx shape than Mojave indigobush.

The specific epithet, *arborescens*, means "treelike," and its thick trunk, thorn-tipped branches, and silvery white bark might remind you of a desert bonsai.

Rushlike Scurf-Pea *Psoralidium junceum*
—Photograph © 2006 by Daniela Roth/Navajo Natural Heritage Program

Mojave Indigobush *Psorothamnus arborescens* —Photograph © 2006 by Regan Dale

Mojave Indigobush *Psorothamnus arborescens* —Photograph by Lori Makarick/National Park Service

Desert Senna *Senna covesii*
PEA FAMILY (Fabaceae/Leguminosae) (*Cassia covesii*)

Plants: Perennial herb, to 50 centimeters tall. **Leaves:** Alternate, pinnately compound, with four to six leaflets, 2 to 7 centimeters long. **Flowers:** In racemes, yellow, to 1.2 centimeters long. **Fruit:** Pod. **Flowering Season:** April to October. **Elevation:** 1,200 to 3,700 feet.

Desert senna, a delicate plant with divided, soft, grayish green leaves, has showy, yellow flowers that are not pea-shaped like the flowers of many species in this family. They lack a banner and keel and have five spoon-shaped petals that are evenly spaced and more or less the same size. In the absence of flowers, you can identify desert senna by its distinct combination of paired, velvety leaflets and erect, light brown pods that remain open on the plant long after the seeds have fallen. The fruits resemble those of other legumes.

Most *Senna* species are found in the tropics, as are many other members of this family. Desert senna occurs at low elevations in New Mexico, Arizona, Nevada, California, and northern Mexico. In the Grand Canyon, it grows in gravelly washes and on talus slopes and sandy beaches from around Cardenas Creek (River Mile 71) downstream to Grand Wash Cliffs.

Senna species have evolved a very small pore at the base of their anther to facilitate "buzz-pollination" by large carpenter bees (*Xylocopa* species) and bumble bees (*Bombus* species). The bees hang upside down and use their wings to vibrate the anther, causing the protein-rich pollen to rain down upon them. The nonsticky pollen does not clog the pore but clings to the bees' fine hairs. The pollen-cloaked bees fly off to the next flower where they rub against the stigmas and pollinate them.

Many plants in the genus *Senna* are called "sensitive plants" because the leaves curl inward in response to touch, which may be a mechanism to protect them from herbivory.

Eyed Gilia *Gilia ophthalmoides*
PHLOX FAMILY (Polemoniaceae)

Plants: Erect annual herb, 15 to 30 centimeters tall. **Leaves:** In basal rosette, pinnately lobed, with dense cobwebby hairs. **Flowers:** Solitary or in clusters, white to purple, tubular, to 5 millimeters long. **Fruit:** Capsule. **Flowering Season:** April to July. **Elevation:** 2,300 to 7,500 feet.

Eyed gilia is so slender it often goes unnoticed, with tiny flowers that range from white to pink to blue. It has a penchant for living in hot, barren, inhospitable places.

Eyed gilia grows from California to Colorado and Arizona. In the Grand Canyon, there are roughly ten species of *Gilia*, and many are very troublesome to tell apart without a microscope. Eyed gilia is one of the more common species, favoring a variety of habitats from the beaches along the Colorado River to open areas in piñon-juniper woodlands and ponderosa pine forests.

The specific epithet, *ophthalmoides*, is Greek for "appearing like the eye."

Desert Senna *Senna covesii*
—Photograph © 2006 by Kathy Darrow

Eyed Gilia *Gilia ophthalmoides*
—Photograph © 2006 by David Inouye

Eyed Gilia *Gilia ophthalmoides* —Photograph © 2006 by Kathy Darrow

Bristly Calico *Langloisia setosissima* ssp. *setosissima*
PHLOX FAMILY (Polemoniaceae)

Plants: Tufted winter annual, to 5 centimeters tall. **Leaves:** Alternate, lanceolate to oblong, with three to five bristly teeth at tip, to 2 centimeters long. **Flowers:** In terminal clusters, pinkish lavender, bristly tipped, to 2 centimeters long. **Fruit:** Capsule. **Flowering Season:** February to June. **Elevation:** 1,200 to 4,500 feet.

In early spring, bristly calico dapples slopes and flats with its pinkish lavender flowers. Although the plant is small, its bristly, five-petaled flowers loom large above the deep green, spine-tipped leaves.

Bristly calico occurs from California to Arizona and south to Mexico. In the Grand Canyon, it grows on talus slopes, flats, and sandy washes the length of the river.

A look-alike species, **Schott's calico (*Loeseliastrum schottii*)**, also has bristly leaves and flowers, but its mostly white flowers have irregular symmetry, appearing to have four petals on top with one on the bottom. Compare these to the radially symmetrical, pinkish lavender flowers of bristly calico.

The genus, *Langloisia*, is named after the Reverend Auguste Barthelemy Langlois (1832–1899), a Louisiana botanist.

Bigelow's Linanthus *Linanthus bigelovii*
PHLOX FAMILY (Polemoniaceae)

Plants: Annual herb, to 20 centimeters tall. **Leaves:** Opposite, narrowly linear, to 3 centimeters long. **Flowers:** In cymes, white to blue. **Fruit:** Capsule. **Flowering Season:** February to May. **Elevation:** 1,200 to 4,400 feet.

While all the flowers on an individual Bigelow's linanthus are the same color, different plants produce a variety of colors, ranging from white to blue. This variation, called *polymorphism*, is a common trait in species of this genus. Researchers investigating this phenomenon have found that reproductive success varies between white- and purple-flowered plants. The five-petaled flowers with long corolla tubes and translucent windows that separate the fused sepals are characteristic of the phlox family.

Bigelow's linanthus occurs from Texas west to California. In the Grand Canyon, it grows from around Nankoweap Creek downstream to Grand Wash Cliffs.

Another closely related annual, **diffuse eriastrum (*Eriastrum diffusum*)**, also has small, white to blue flowers, but it differs in having alternate leaves and densely woolly flowering stalks. These species have overlapping ranges but diffuse eriastrum more commonly grows in the western portion of the Grand Canyon.

The specific epithet honors John Milton Bigelow (1804–1878), a professor of botany who collected plants as part of the Pacific Railroad Survey of 1853–1854.

Bristly Calico *Langloisia setosissima* ssp. *setosissima*
—Photograph © 2006 by Kristin Huisinga

Diffuse Eriastrum *Eriastrum diffusum*
—Photograph © 2006 by Max Licher

Bigelow's Linanthus *Linanthus bigelovii* —Photograph © 2006 by Kathy Darrow

Lanceleaf Plantain
Plantago lanceolata

PLANTAIN FAMILY (Plantaginaceae)

Plants: Exotic perennial herb, to 50 centimeters tall. **Leaves:** In basal rosettes, lanceolate to oblanceolate, with prominent veins, 3 to 25 centimeters long. **Flowers:** In terminal spikes that are up to 8 centimeters long; flowers inconspicuous. **Fruit:** Capsule. **Flowering Season:** May to September. **Elevation:** 2,800 to 8,200 feet.

Lanceleaf plantain, a hearty perennial, has broad, succulent, deep green leaves and thick roots that store nutrients even during the harshest times of year. It thrives along riverbanks and side creeks in gravelly areas where floods bring new soils and occasional standing water.

Common plantain (*Plantago major*), another perennial, has broad, oval-shaped leaves that are arranged in a flattened rosette. In contrast, the leaves of lanceleaf plantain are narrower and stand in an erect rosette. Both have tall flowering stalks that bear clusters of tiny, four-petaled flowers. Look closely to see that the sepals are fused in lanceleaf plantain and free in common plantain.

These perennial species are European immigrants and now occur worldwide. In the Grand Canyon, they grow the length of the river and to both rims, favoring moist soils.

The genus name, *Plantago*, comes from the Latin word *planta*, meaning "footprint." These plants are known as white man's foot, referring to their abundance in human-disturbed areas. The seeds are used for medicine and food, and the leaves of common plantain are a nutritious wild food eaten raw or boiled.

Woolly Plantain
Plantago patagonica
(*Plantago purshii*)

PLANTAIN FAMILY (Plantaginaceae)

Plants: Annual herb, to 38 centimeters tall. **Leaves:** In basal rosettes, linear to linear-lanceolate, hairy, 1 to 15 centimeters long. **Flowers:** In terminal spikes that are up to 13 centimeters long; flowers inconspicuous. **Fruit:** Capsule. **Flowering Season:** January to August. **Elevation:** 1,200 to 7,000 feet.

Woolly plantain, a dainty winter annual, blankets slopes with a grayish green hue. Lacking adequate winter and spring rain, the seeds of these plants stay dormant in the soil, waiting for more ideal conditions.

Woolly plantain occurs from Canada south to California and Texas and into Mexico. It also lives in Patagonia, Chile—hence its specific epithet. In the Grand Canyon, it grows on desert slopes and flats and in drainages the length of the canyon.

Another annual with an overlapping range, **blonde plantain (*Plantago ovata*)**, has a very similar morphology. Compare the membranous, egg-shaped bracts of blonde plantain to the triangular flower bracts that conspicuously protrude from the dense flowering spike of woolly plantain.

Both annual plantains provide important food for mammals, especially during early spring when little else is available. Many native people add the ground seeds to a corn drink called *pinole*. The seeds contain high amounts of mucilage, which lubricates the digestive tract and buffers the uptake of sugar, thus helping to regulate diabetes.

Lanceleaf Plantain *Plantago lanceolata*
—Photograph © 2006 by Geoff Gourley

Common Plantain *Plantago major*
—Photograph © 2006 by Kristin Huisinga

Woolly Plantain *Plantago patagonica*
—Photograph © 2006 by Kate Watters

Blonde Plantain *Plantago ovata*
—Photograph © 2006 by Kristin Huisinga

Prickly-Poppy   *Argemone munita*
POPPY FAMILY (Papaveraceae)

Plants: Perennial herb, to 1.5 meters tall. **Leaves:** Alternate, pinnately lobed, very prickly, clasping at base, 3 to 15 centimeters long. **Flowers:** Solitary, white, to 13 centimeters wide. **Fruit:** Prickly capsule. **Flowering Season:** April to October. **Elevation:** 3,000 to 7,200 feet.

The delicate, papery petals of prickly-poppy welcome pollinators of all sorts, while its well-armed leaves and stems deter herbivores. At first glance, the bluish green spiny leaves of this plant may remind you of red barberry (*Mahonia haematocarpa*) or New Mexico thistle (*Cirsium neomexicanum*). To confirm the identity of prickly-poppy when it lacks flowers, scar a small part of the stem and look for its thick, yellow or orange sap.

Prickly-poppy occurs in California and Nevada east to New Mexico, and south to Mexico. In the Grand Canyon, it most commonly grows at higher elevations, but look for it at river level near Rider and North Canyons.

A related species, **Roaring Springs prickly-poppy (***Argemone arizonica***)** is endemic to the Grand Canyon, known only from along the North Kaibab Trail. It resembles prickly-poppy, but Roaring Springs prickly-poppy has prickles only on its leaf veins and sparsely on its stems. Prickly-poppy has prickles covering its entire leaf and stem surface. Another member of the poppy family in the Grand Canyon is **little gold poppy (***Eschscholzia minutiflora***)**. It has four-petaled, orange flowers and highly divided leaves.

The leaves, pulverized seeds, and sap of prickly-poppy are administered externally for warts, rashes, hives, and abrasions. It is mildly narcotic, so it is used as a pain reliever and sedative, but it is toxic in large quantities. Prickly-poppy can become weedy in disturbed areas and is a problem on overgrazed lands because livestock will not eat it.

Cave Primrose   *Primula specuicola*
PRIMROSE FAMILY (Primulaceae)   (*Primula hunnewellii*)

Plants: Perennial herb, to 28 centimeters tall. **Leaves:** Basal, spatulate, toothed, strongly white-mealy beneath, 2 to 20 centimeters long. **Flowers:** In umbels, white to pink. **Fruit:** Capsule. **Flowering Season:** March to May. **Elevation:** 1,250 to 7,700 feet.

Cave primrose, one of the most showy hanging garden species on the Colorado Plateau, is endemic to the canyons of the Colorado River watershed, where it frequents sandstone alcoves. While many hanging garden species flower during the summer monsoons, cave primrose flowers early in the spring. In the Grand Canyon, it grows at side canyon seeps and springs and is known only from Buck Farm Canyon and Bert's Canyon (just downstream from Buck Farm Canyon) in the river corridor. However, it is more common at high-elevation sites in the Grand Canyon and in Glen Canyon National Recreation Area.

The genus name, *Primula*, is Latin for "little firstling" because many are spring bloomers. The specific epithet, *specuicola*, means "cave-dwelling," referring to its preference for hanging gardens in alcoves.

Prickly-Poppy *Argemone munita* —Photograph © 2006 by David Inouye

Little Gold Poppy *Eschscholzia minutiflora* —Photograph © 2006 by Kate Watters

Cave Primrose *Primula specuicola*
—Photograph © 2006 by Al Schneider/www.swcoloradowildflowers.com
Inset —Photograph © 2006 by Al Schneider/www.swcoloradowildflowers.com

Krameria, Littleleaf Ratany
RATANY FAMILY (Krameriaceae)

Krameria erecta
(*K. parvifolia*)

Plants: Shrub, to 1 meter tall. **Leaves:** Alternate, lanceolate, spine-tipped, to 2 centimeters long. **Flowers:** Solitary in axils, deep pink to purple. **Fruit:** Heart-shaped pod. **Flowering Season:** April to October. **Elevation:** 1,400 to 4,200 feet.

Krameria, one of the most interesting yet least-known shrubs in the Grand Canyon, is often overlooked when not in bloom because its leaves are small and a rather subdued gray green. Its deep burgundy flowers, however, are hard to miss. Because krameria is a poor photosynthesizer, its roots attach to the roots of other plants to appropriate nutrients. You will see it growing among its host plants such as Mormon tea (*Ephedra* species). As if evolution wanted to even the score, a tiny parasitic moth larvae inhabits krameria's fruit and eats the seeds, rendering most of them inviable. Look for the larvae's tiny escape hole as you examine the prickly fruits.

Many flowers attract pollinators through visual or fragrant cues and offer a reward of nectar, a quick-energy fluid that encourages return visitation. Krameria flowers, however, produce oil, a richer energy source. Its primary pollinator, the digger bee (*Centris* species), collects the oil with its squeegeelike hind legs and feeds it to its larva.

Krameria grows in creosotebush and blackbrush communities from California east to Texas and south to Mexico. In the Grand Canyon, it inhabits slopes and benches downstream from Whitmore Trail.

Native people used krameria to make a deep red dye and also to treat hemorrhoids, sore throats, dental problems, and sore nipples.

Blackbrush
ROSE FAMILY (Rosaceae)

Coleogyne ramosissima

Plants: Compact shrub, to 1.5 meters tall. **Leaves:** Opposite, clustered, linear, involute margins, to 1.5 centimeters long. **Flowers:** Solitary, yellow, to 6.5 millimeters long. **Fruit:** Achene. **Flowering Season:** March to May. **Elevation:** 2,900 to 5,600 feet.

From the rims of the Grand Canyon, blackbrush appears as small gray clumps dominating the vast landscape of the Tonto Plateau. It is the only species in the genus *Coleogyne* in the United States and occurs over a wide geographic range in California, Nevada, Colorado, Utah, and Arizona, commonly inhabiting the transition zone between warm and cold deserts. In the Grand Canyon, the Tonto Plateau marks the transitional zone between the warmest part of the Great Basin Desert and the coolest portion of the Mojave Desert. This is where blackbrush rules, a dominance that gives the plant community the name *blackbrush community*.

Blackbrush can live for well over a century, with very few young plants in a stand. When rains coax out the petal-less blossoms with four yellow sepals, the entire population flowers and produces seed within a few weeks. The many intersecting, alternate branches bear sharp tips, providing an intricate shelter for many insect, ant, and spider species. The greenery also provides important cover and food for numerous bird and mammal species, but it has low nutritional value.

Krameria *Krameria erecta* —Photograph © 2006 by Kathy Darrow
Flower Inset —Photograph © 2006 by Kathy Darrow
Top Inset —Photograph © 2006 by Kathy Darrow

Blackbrush *Coleogyne ramosissima* —Photograph © 2006 by Kate Thompson
Inset —Photograph © 2006 by Kate Watters

Apache Plume   *Fallugia paradoxa*
ROSE FAMILY (Rosaceae)

Plants: Shrub, to 2 meters tall. **Leaves:** Alternate, pinnately lobed, deep green, to 2 centimeters long. **Flowers:** Solitary, white, 3 to 5 centimeters wide. **Fruit:** Feathery achenes. **Flowering Season:** April to November. **Elevation:** 2,800 to 7,200 feet.

The rose-colored, feathery plumes of the fruits are the most striking feature of Apache plume, inspiring someone to name the plant after Native American war regalia, which was adorned with many plumose feathers. Downstream of Lees Ferry, scan the slopes for the distinct band of Apache plume near the pre-dam high-water mark.

The fruits easily distinguish Apache plume from its relatives, mountain mahogany (*Cercocarpus montanus*) and cliffrose (*Purshia stansburiana*). Compared with the twenty or more long tails of Apache plume fruits, mountain mahogany fruits have only one tail and cliffrose fruits have four to ten, much shorter tails. In the absence of fruit, note the bleached white bark and shorter, branching habit of Apache plume.

Apache plume grows abundantly throughout the West, reaching its lowest elevation in the Grand Canyon where it is a rare find along the river downstream of Nankoweap Creek. It thrives in side canyons like Deer Creek and many others.

Apache plume branches were used as brooms, ladder-back cradleboard rungs, and arrow shafts. The genus name commemorates the seventeenth-century Italian abbot and botanical writer, Virgilio Fallugi (?–1707).

Rockmat, Mat Rockspirea   *Petrophyton caespitosum*
ROSE FAMILY (Rosaceae)

Plants: Prostrate shrub, to 1 meter wide. **Leaves:** Alternate, spatulate, gray green, with dense, silky hairs, 3 to 17 millimeters long. **Flowers:** In compact spikes that are 2 to 14 centimeters long; flowers white, tiny. **Fruit:** Dry follicle. **Flowering Season:** June to October. **Elevation:** 1,200 to 7,800 feet.

Rockmat forms dense grayish green masses near seeps and springs, often clinging to limestone ledges or draping over cliff edges. The woody branches creep horizontally and the small flowering spikes rise above the clustered leaves in the late summer.

Rockmat occurs from Oregon and California east to South Dakota and Texas, preferring limestone substrates and seeps rich in calcium carbonate. In the Grand Canyon, it grows on both the North and South Rims and in hanging gardens throughout the Inner Canyon. It is especially conspicuous at Elves Chasm and Fern Glen Canyon.

The roots are used medicinally. The genus name comes from the Greek words *petros*, meaning "rock," and *phyton*, meaning "plant."

Apache Plume *Fallugia paradoxa* —Photograph © 2006 by David Inouye
Inset —Photograph © 2006 by David Inouye

Fruit of Apache Plume *Fallugia paradoxa*
—Photograph © 2006 by Glenn Rink

Rockmat *Petrophyton caespitosum*
—Photograph © 2006 by Roger Dale

Wild Rose, Arizona Rose *Rosa woodsii* var. *ultramontana*
ROSE FAMILY (Rosaceae) (*Rosa arizonica*)

Plants: Deciduous shrub, to 5 meters tall. **Leaves:** Alternate, pinnately compound, with three to seven toothed leaflets, 1 to 3 centimeters long. **Flowers:** Solitary, pale pink. **Fruit:** Berrylike hips. **Flowering Season:** April to August. **Elevation:** 3,500 to 8,500 feet.

The captivating fragrance of wild rose surpasses that of most cultivated roses. The sharp, curved prickles on its stems and its thicket-forming habit make it easy to differentiate from other shrubs, even without flowers.

Wild rose flourishes in many different habitats from New Mexico west to California and north to Washington, Idaho, and Montana. In the Grand Canyon, it blooms in shady locations at higher elevations and in side canyons such as Roaring Springs (off the North Kaibab Trail) and Havasu Canyon, so hikers are more likely to find wild rose than river travelers. The Grand Canyon harbors a rare endemic species, **desert rose (*Rosa stellata* ssp. *abyssa*)**. It is smaller, has much pricklier branches and fruits, and grows at higher elevations.

Rose seeds are hard and resilient to the digestive juices of animals, whose droppings provide a nice location for germination. Collected in midwinter after they have frozen and thawed, the softer hips are a good source of vitamin C. The fruits can also be made into wines, jams, and jellies. The genus name commemorates Joseph Woods (1776–1864), an English architect and botanical author.

Common Paintbrush *Castilleja applegatei* ssp. *martinii*
SNAPDRAGON FAMILY (Scrophulariaceae) (*Castilleja chromosa*)

Plants: Perennial herb, to 55 centimeters tall. **Leaves:** Alternate, linear to lanceolate, green to purple, lobed, to 6 centimeters long. **Flowers:** In terminal spikes, tubular, reddish orange, 2 to 4 centimeters long. **Fruit:** Capsule. **Flowering Season:** March to August. **Elevation:** 1,700 to 7,800 feet.

Common paintbrush flowers stand out in the desert as if a wandering artist had dipped them in vermillion paint. The bright reddish orange "petals" are actually sepals and bracts, which tightly enclose the green-tipped, tubular petals. As a seedling, common paintbrush initially relies on its own resources for nutrients and water. As it grows, it becomes a partial parasite, seeking out and penetrating the roots of other plants for nourishment. The often reddish purple of its leaves and stems suggests that it does not rely completely on photosynthesis.

Common paintbrush is the most widespread of more than forty different species of paintbrush found throughout western North America. In the Grand Canyon, it tolerates hot, dry, sandy soils and talus slopes the length of the river, extending to the rims.

Another species, **long-leaved paintbrush (*Castilleja linariifolia*)**, has longer, unlobed leaves and is often more robust because it grows near the river's edge. Its flaming blossoms extend into October. Overlapping in range with common paintbrush, it also finds a home on plateaus and side canyons throughout the Inner Canyon up to the rims.

Common paintbrush provides an early nectar source for hummingbirds during their spring migrations. Paintbrush can be made into a medicinal tea to soothe upset stomachs, cramps, and painful bites. The genus name, *Castilleja*, pays homage to Domingo Castillejo (1744–1793), a Spanish botany professor.

Wild Rose *Rosa woodsii* var. *ultramontana* —Photograph © 2006 by Kate Watters
Inset —Photograph © 2006 by Max Licher

Common Paintbrush *Castilleja applegatei* ssp. *martinii* —Photograph © 2006 by David Edwards

Twining Snapdragon
Maurandella antirrhiniflora
SNAPDRAGON FAMILY (Scrophulariaceae)
(*Maurandya antirrhiniflora*)

Plants: Vinelike perennial herb, to 80 centimeters long. **Leaves:** Alternate, triangular to hastate, shallowly lobed, to 3 centimeters long. **Flowers:** Solitary in axils, pink to purple, to 3 centimeters long. **Fruit:** Capsule. **Flowering Season:** April to October. **Elevation:** 1,200 to 6,300 feet.

Twining snapdragon twists and twines around anything it encounters. Its magenta blossoms resemble garden-variety snapdragon flowers but are much smaller. The petals are fused into a swollen, closed tube crowned with two lips, termed *personate* by botanists because the petals look like human lips pursed and awaiting a kiss. A consistent water supply will yield nearly year-round blossoms. The bright yellow palate on the lower lip of the flower beckons bees to crawl into the tube, where they get dusted with fresh pollen. Some flowers have evolved to fertilize themselves without opening, further assuring successful reproduction.

Twining milkweed (*Sarcostemma cynanchoides* ssp. *cynanchoides*) is another plant with a vinelike habit and arrow-shaped leaves. However, its cluster of white to pink flowers and odoriferous, milky sap make it unmistakable.

Today, twining snapdragon ranges from Texas west to Nevada. It is now rare in California and Utah, where it has been little documented since the 1870s. In the Grand Canyon, it grows among shrubs in rocky washes and on slopes the length of the river. You are most likely to find this beauty on hikes in side canyons with perennial water.

This genus takes its name for Catalina Pancracia Maurandy (n.d.), an eighteenth-century Spanish botany professor.

Cardinal Monkeyflower
Mimulus cardinalis
SNAPDRAGON FAMILY (Scrophulariaceae)

Plants: Rhizomatous perennial herb, to 1 meter tall. **Leaves:** Alternate, ovate to obovate, irregularly toothed, 2 to 10 centimeters long. **Flowers:** Solitary, orange red to scarlet, to 5 centimeters long. **Fruit:** Capsule. **Flowering Season:** March to November. **Elevation:** 1,200 to 6,000 feet.

Cardinal monkeyflower, a ruby-throated showstopper, forms dense patches with bright red flowers that contrast against dark green leaves and water-worn rocks. In hanging gardens and seeps, its widely ovate leaves blend in with its associates, helleborine orchid (*Epipactis gigantea*) and maidenhair fern (*Adiantum capillus-veneris*). Cardinal monkeyflower resembles another red-flowered, water-loving plant, scarlet lobelia (*Lobelia cardinalis*). Scarlet lobelia has milky sap and narrower, more distinctly separated petals.

Cardinal monkeyflower occurs from Oregon south to Baja California and east through Mexico into New Mexico. In the Grand Canyon, it prospers in shady areas near seeps and springs and along the edges of gently flowing perennial streams in side canyons. It grows from Vasey's Paradise downstream to Grand Wash Cliffs.

The Grand Canyon hosts seven species of *Mimulus*, four of which occur in the lower elevations of the Inner Canyon and are mostly yellow-flowered. The common **yellow monkeyflower (*M. guttatus*)** is the most widespread monkeyflower species in Arizona. Without flowers, yellow monkeyflower and cardinal monkeyflower are virtually indistinguishable.

Top: **Cardinal Monkeyflower** *Mimulus cardinalis* —Photograph © 2006 by David Inouye
Bottom: **Yellow Monkeyflower** *Mimulus guttatus* —Photograph © 2006 by John Running

Cardinal Monkeyflower *Mimulus cardinalis* —Photograph © 2006 by Kathy Darrow

Twining Snapdragon *Maurandella antirrhiniflora* —Photograph © 2006 by Kristin Huisinga

Palmer's Penstemon
Penstemon palmeri
SNAPDRAGON FAMILY (Scrophulariaceae)

Plants: Perennial herb, to 1.5 meters tall. **Leaves:** Opposite, lanceolate to ovate, serrate, sessile, blue green, to 20 centimeters long. **Flowers:** In contracted panicles, creamy white to pink, to 4 centimeters long. **Fruit:** Capsule. **Flowering Season:** March to September. **Elevation:** 1,400 to 7,000 feet.

The inflated, soft pink blossoms of Palmer's penstemon decorate tall stalks and invite a medley of pollinators. A yellow, bearded stigma protrudes from the flower's throat, poised for a dusting of pollen. In the absence of flowers, look for the stout, opposite, clasping leaves that become fused near the top of the plant, appearing as one elongated leaf.

Palmer's penstemon occurs in California, Arizona, Utah, and Nevada. In the Grand Canyon, it is uncommon but can be found in side canyons and along trails the length of the river, extending to both rims.

Another penstemon species with overlapping distribution, **Eaton's firecracker (*Penstemon eatonii*)** is more abundant at higher elevations, especially along trails. It has striking, narrowly tubular, orangish red to bright red blossoms and untoothed, deep green leaves.

The genus *Penstemon* is prolific throughout the West. A single Palmer's penstemon plant produces thousands of tiny seeds that are easily collected, leading to its use in seed mixes. The specific epithet, *palmeri*, credits Ernest Jesse Palmer (1875–1962), a poet, author, and self-taught botanist, who has over two hundred plant species named in his honor.

Water Speedwell, Veronica
Veronica anagallis-aquatica
SNAPDRAGON FAMILY (Scrophulariaceae)

Plants: Exotic, semiaquatic, perennial herb, to 1 meter tall. **Leaves:** Opposite, lanceolate to ovate, toothed, to 8 centimeters long. **Flowers:** Solitary or in racemes, sky blue to purple, to 1 centimeter long. **Fruit:** Heart-shaped capsule. **Flowering Season:** March to October. **Elevation:** 1,200 to 3,200 feet.

The tiny, sky blue flowers of water speedwell have unequal petals and two delicate stamens covered with pollen, which butterflies inadvertently spread when they drink nectar. This fast-growing, rhizomatous plant forms dense patches, and its creeping, reddish stems are often exposed at the water's edge.

Water speedwell is a native of Europe, Africa, and Asia and is now widely distributed throughout North America. In the Grand Canyon, you will discover water speedwell in moist soils on riverbanks and near springs and side streams the length of the river.

A native species in the Grand Canyon, **American brooklime (*Veronica americana*)** has leaves with petioles while those of water speedwell clasp the stem. While overlapping in range with water speedwell at river level, American brooklime finds its way into moist areas extending to the North Rim.

In most species of the genus, the aerial parts are edible raw or cooked and contain vitamin C. The genus name refers to the markings on some species that are said to resemble the handkerchief that St. Veronica offered to Jesus on his way to Calvary.

Palmer's Penstemon *Penstemon palmeri*
—Photograph © 2006 by Lisa A. Hahn

Palmer's Penstemon *Penstemon palmeri*
—Photograph © 2006 by Lisa A. Hahn

Eaton's Firecracker *Penstemon eatonii*
—Photograph © 2006 by Kate Watters

Water Speedwell *Veronica anaga'lis-aquatica* —Photograph © 2006 by Kate Watters

Inset —Photograph © 2006 by Tina Ayers

Fendler's Sandmat *Chamaesyce fendleri*
SPURGE FAMILY (Euphorbiaceae) (*Euphorbia fendleri*)

Plants: Matted perennial herb, 5 to 15 centimeters long. **Leaves:** Opposite, oval to oblong, 3 to 11 millimeters long. **Flowers:** One per node, 1.5 millimeters diameter. **Fruit:** Four-angled capsule. **Flowering Season:** May to September. **Elevation:** 1,450 to 7,400 feet.

Fender's sandmat, an inconspicuous member of the Grand Canyon's flora, has a unique flower structure: white, petal-like male flowers surround a central, stalked female flower. This dainty red-tinged plant bleeds milky sap from its leaves and stems, like other members of the spurge family.

Fendler's sandmat occurs from California east to Nebraska and south to Texas and Mexico. In the Grand Canyon, it grows in rocky and sandy soils from the river to the rims.

The specific epithet, *fendleri*, commemorates Augustus Fendler (1813–1883), a German plant collector in North and Central America. In 1846, Fendler joined a contingent of American soldiers in New Mexico, where he collected more than a thousand specimens during a one-year period, all within walking distance of Santa Fe.

Ross's Spurge, Marble Canyon Spurge *Euphorbia aaron-rossii*
SPURGE FAMILY (Euphorbiaceae)

Plants: Perennial herb to small shrub, to 1 meter tall. **Leaves:** Alternate, lanceolate, 2 to 3 centimeters long. **Flowers:** Solitary, pinkish white to greenish, tiny. **Fruit:** Four-angled capsule. **Flowering Season:** April to October. **Elevation:** 2,100 to 3,300 feet.

Ross's spurge appears wiry and often strawlike because it is mostly leafless. Its tiny pinkish white flowers crown the tips of its slender branches. When the stems are broken, it exudes a toxic, white latex, which is characteristic of the spurge family. Ross's spurge is only known from the Grand Canyon and grows on rocky talus slopes and sandy soils of old river bars and dunes in Marble Canyon and its side canyons.

Ross's spurge is a recent addition to the Arizona flora, as it was not described until 1988. It was named for its discoverer, Aaron B. Ross (1917–1973), a physician from Utah who was also a naturalist and boatman.

Fendler's Sandmat *Chamaesyce fendleri* —Photograph © 2006 by Lisa A. Hahn

Ross's Spurge *Euphorbia aaron-rossii*
—Photograph © 2006 by Lisa A. Hahn

Ross's Spurge *Euphorbia aaron-rossii*
—Photograph © 2006 by Lisa A. Hahn

Desert Rock Nettle, Velcro Plant *Eucnide urens*
STICKLEAF FAMILY (Loasaceae)

Plants: Perennial herb, to 1 meter tall. **Leaves:** Alternate, ovate to round, coarsely toothed, 1 to 7 centimeters long. **Flowers:** In terminal cymes, cream to pale yellow, 2 to 5.5 centimeters long. **Fruit:** Prickly capsule. **Flowering Season:** April to August. **Elevation:** 1,200 to 3,000 feet.

The large, waxy, translucent flowers of desert rock nettle are appealing from a distance but will punish those who pick them. The entire plant cloaks itself in stinging hairs in an attempt to defend itself against animals that try to eat its seemingly luscious leaves. The barbed hairs, which penetrate and hook your skin, are so formidable that bats have reportedly tangled themselves on plants growing near cave entrances.

Desert rock nettle occurs in warm desert communities in California, Nevada, Utah, Arizona, and northwest Mexico. In the Grand Canyon, it lives in rocky crevices on cliff faces in side canyons and dry rocky slopes from the Bright Angel Trail downstream to Grand Wash Cliffs.

Oddly enough, desert bighorn sheep occasionally browse this plant. The genus name, *Eucnide*, is derived from the Greek words *eu*, meaning "good" or "pretty," and *knide*, meaning "stinging nettle," descriptive of this strongly nettlelike plant. The specific epithet, *urens*, translates as "stinging" or "burning."

Blazing Star, Stickleaf *Mentzelia pumila*
STICKLEAF FAMILY (Loasaceae)

Plants: Biennial or short-lived perennial herb, to 50 centimeters tall. **Leaves:** In rosettes, lanceolate to oblong, dark green, to 10 centimeters long. **Flowers:** Solitary or in cymes; yellow, to 1.5 centimeters long. **Fruit:** Narrowly winged capsule. **Flowering Season:** March to September. **Elevation:** 2,000 to 4,500 feet.

Blazing star, with its striking, star-shaped, lemon yellow flowers and crowds of anthers dusted with pollen, opens in the late afternoon and closes around sunset. The flowers appear to have ten petals, but the larger five are actually sepals, with their inner surface bearing the same color as the petals. Even without the flower, you can identify this plant by its sandpapery, dark green leaves with wavy margins, ghostly white stems, and starlike whorl crowning the bullet-shaped seed capsule. The barbed hairs of blazing star give it the distinct ability to stick to clothing like Velcro, leading some to call it the boutonniere plant.

There are at least seven other *Mentzelia* species found in the Grand Canyon, but blazing star is the most common. It has a wide distribution, growing from Montana and North Dakota south to Colorado and New Mexico and west to Utah. In the Grand Canyon, you will see it on dry slopes and in rocky washes from Nankoweap Creek downstream to Grand Wash Cliffs.

You might confuse blazing star with desert rock nettle (*Eucnide urens*) because both have sticky leaves, star-shaped flowers, and an affinity for rocky washes. The broader, oval-shaped leaves of rock nettle set this species apart from the lance-shaped, lobed leaves of blazing star.

Native people roast or parch the seeds and grind them into a flour to eat as mush, bread, or gravy. Some use the dried ground root as a laxative. The genus name honors Christian Mentzel (1622–1701), a seventeenth-century German botanist and physician.

Desert Rock Nettle *Eucnide urens* —Photograph © 2006 by Roger Dale

Desert Rock Nettle *Eucnide urens*
—Photograph © 2006 by Lisa A. Hahn

Blazing Star *Mentzelia pumila*
—Photograph © 2006 by Joe Pollock

Skunkbush
Rhus trilobata var. *simplicifolia*

SUMAC FAMILY (Anacardiaceae)

Plants: Shrub, to 2.5 meters tall. **Leaves:** Alternate, round, shallowly lobed, dark green, to 2.5 centimeters long. **Flowers:** In dense, terminal spikes, light yellow. **Fruit:** Red drupe. **Flowering Season:** April to May. **Elevation:** 1,400 to 6,500 feet.

The distinct odor of citrus may be your first clue to identify skunkbush. In the spring, its compact, yellow, nonshowy flowering clusters appear before its shiny, dark green leaves emerge. The leaves turn bright red in the fall. When in season, the sticky, red berries also help identify this plant. Glandular hairs cover the berry and impart a sweet-sour taste, giving the plant another common name, *lemonade-berry*.

Several varieties of skunkbush occur throughout western North America. In the Grand Canyon, our most common variety (*Rhus trilobata* var. *simplicifolia*) grows along the river and in side canyons, usually confined to shaded, cool slopes at lower elevations. You can also spot skunkbush in unlikely places such as debris fans at the mouths of side canyons where seeds have washed down from above.

Rhus trilobata* var. *trilobata has leaves divided into three parts and is more common at higher elevations. Also in this genus is the much less common and altogether different **smooth sumac (R. glabra)**, which has much larger, pinnately compound leaves. Look for it at Stone Creek and on the North Kaibab Trail where its leaves blaze red in the fall.

The branches provide a flexible, durable basketry material and the berries make a tart, tasty lemonade. Because skunkbush is closely related to poison-ivy, those who have a severe susceptibility to the latter may also develop dermatitis through skin contact with skunkbush's sap.

Poison-Ivy
Toxicodendron rydbergii

SUMAC FAMILY (Anacardiaceae)

Plants: Perennial herb or small shrub, to 30 centimeters tall. **Leaves:** Alternate, with three toothed leaflets; leaves to 11 centimeters long. **Flowers:** In axillary racemes or panicles, white to greenish yellow, inconspicuous. **Fruit:** Berry. **Flowering Season:** April to September. **Elevation:** 1,900 to 3,000 feet.

Poison-ivy causes severe dermatitis. Although the leaves of this plant drop in the winter, contact with the sap can still cause a strong reaction. The little ditty "leaves three, let it be; berries white, take flight" is a handy mnemonic device for remembering its features. When young, the three-parted leaves of Arizona boxelder (*Acer negundo* var. *arizonicum*) resemble poison-ivy, but these are opposite on the stem while those of poison-ivy are alternate.

Poison-ivy occurs throughout much of temperate North America but is not adapted to dry sites. In the Grand Canyon, it grows only in a few permanently moist areas with springs or seeps. Best known of these are Vasey's Paradise and Deer Creek.

The genus name, *Toxicodendron*, means poison tree. The specific epithet, *rydbergii*, honors Per Axel Rydberg (1860–1931), the first curator of the herbarium at the New York Botanical Garden and author of the first flora of the Rocky Mountains.

Skunkbush *Rhus trilobata* var. *simplicifolia* —Photograph © 2006 by Daniela Roth/Navajo Natural Heritage Program

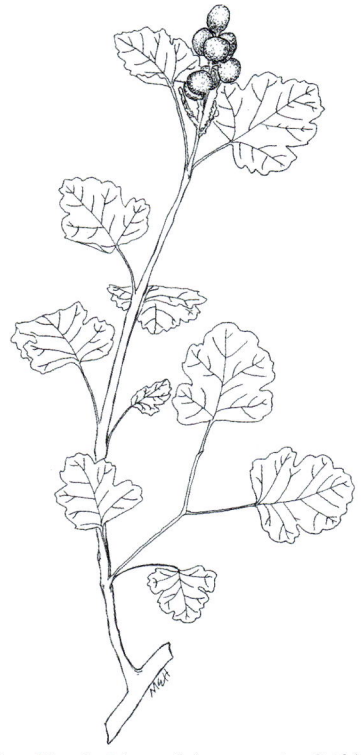

Skunkbush *Rhus trilobata* var. *simplicifolia*
—Illustration © 2006 by Mar-Elise Hill

Poison-Ivy *Toxicodendron rydbergii*
—Photograph © 2006 by John Running

Brownfoot, Buffalo Fur
SUNFLOWER FAMILY (Asteraceae/Compositae)

Acourtia wrightii
(*Perezia wrightii*)

Plants: Erect perennial herb, to 1.2 meters tall. **Leaves:** Alternate, lanceolate to ovate, sharp-toothed margins, clasping base, 6 to 13 centimeters long, leathery. **Flowers:** In terminal panicles; heads 1 to 1.5 centimeters long; disk flowers lavender pink; ray flowers lacking. **Fruit:** Achene with white bristly pappus. **Flowering Season:** January to October. **Elevation:** 1,200 to 6,000 feet.

Brownfoot is easy to spot even when the attractive, sweet-smelling flowers are past their prime. The clasping leaves dry to a golden brown without withering and stay attached to the stems above a furry root crown, which inspired the common names *brownfoot* and *buffalo fur*. The rust-colored tufts of the root crown are usually hidden beneath the soil, but reveal themselves on steep eroding slopes, such as those on the hike to Saddle Canyon.

Brownfoot is usually associated with desertscrub or piñon-juniper communities in the Southwest, growing from Nevada and Arizona east to Texas, where it was first collected. In the Grand Canyon, brownfoot is widespread on shaded slopes along the river and in side canyons where it often finds shelter beneath the canopy of taller shrubs and trees.

The tufts on the root crown, which resemble cotton, can be applied to bleeding wounds. The specific epithet, *wrightii*, refers to the nineteenth-century plant collector, Charles Wright (1811–1885), who sent many newly discovered plants to Asa Gray, then curator of the herbarium at Harvard. Wright was a teacher in the territory of Texas but his first passion was botany and he literally walked across Texas collecting plants.

White Bursage
SUNFLOWER FAMILY (Asteraceae/Compositae)

Ambrosia dumosa
(*Franseria dumosa*)

Plants: Grayish white perennial shrub, 20 to 60 centimeters tall. **Leaves:** Clustered, ovate, twice pinnately lobed, rounded margins, densely hairy, 1 to 4 centimeters long. **Flowers:** In inconspicuous nodding spikes; male and female heads separate, disk flowers greenish yellow; ray flowers lacking. **Fruit:** Spiny bur. **Flowering Season:** March to November. **Elevation:** 1,200 to 4,000 feet.

White bursage, a compact, drought-tolerant shrub, appears parched during dry periods, when it sheds its leaves to conserve resources. New leaves and flowers emerge following rainy periods, a time when this plant is most recognizable. It acts as a nurse plant, providing shelter for the seedlings of plants such as the pincushion cactus (*Mammillaria* species) and creosotebush (*Larrea tridentata* var. *tridentata*).

White bursage is sometimes confused with big sagebrush (*Artemisia tridentata*) or fourwing saltbush (*Atriplex canescens*) because they are all shrubs with silvery leaves. White bursage is distinctly smaller, with twice-lobed leaves and spiny burs tucked into its leaf axils. While noticeably fragrant, the leaves of white bursage are not salty like those of saltbush, nor do they smell pungent like big sagebrush.

White bursage often forms the dominant understory in the warmer, drier portions of the Sonoran and Mojave Deserts in Arizona, California, Nevada, Utah, and northern Mexico. In the Grand Canyon, white bursage grows from Stone Creek downstream to Grand Wash Cliffs, where it covers large expanses on dry, open slopes.

Annual burweed (*Ambrosia acanthocarpa*), an annual from the same genus, frequents beaches and side canyons in the Grand Canyon. It is easy to distinguish the shrubby white bursage from this slender, bright green herb.

Wildlife and livestock, particularly burros and sheep, browse white bursage. A few members of this genus, known as "ragweed," contribute to the discomfort of allergy sufferers.

Brownfoot *Acourtia wrightii* —Photograph © 2006 by David Inouye

White Bursage *Ambrosia dumosa* —Photograph © 2006 by Kathy Darrow

White Bursage *Ambrosia dumosa* —Photograph © 2006 by Kathy Darrow

Louisiana Wormwood, Water-Sage *Artemisia ludoviciana*
SUNFLOWER FAMILY (Asteraceae/Compositae)

Plants: Perennial herb, 30 to 100 centimeters tall. **Leaves:** Linear to lanceolate, simple to lobed, green gray, 1 to 10 centimeters long. **Flowers:** In panicles; disk flowers inconspicuous, yellowish; ray flowers lacking. **Fruit:** Achene. **Flowering Season:** July to December. **Elevation:** 1,200 to 7,500 feet.

Louisiana wormwood has a strong sagelike scent when you crush its greenish gray herbage. Glandular hairs clothe the leaves and house fragrant essential oils. You can easily spot this plant along the shoreline because the softer, greenish gray leaves are in striking contrast to the other verdant foliage.

Because it has the ability to live in many vegetation zones, Louisiana wormwood is one of the most widely distributed species in western North America. In the Grand Canyon, it grows from river to rim. At lower elevations, it occurs at the water's edge, on rocky slopes, and in side canyons the length of the river.

Other *Artemisia* species found in the Grand Canyon include **big sagebrush (*A. tridentata*)**, a large shrub with leaves that have three distinct teeth at the tips. While dominant in the West, it mostly occurs at higher elevations in the Grand Canyon. **Fringed sage (*A. frigida*)** most closely resembles Louisiana wormwood but it has more minutely divided leaves that are compactly arranged on the stems as compared to the more open leaf arrangement of Louisiana wormwood. Fringed sage is only documented at two locations in the lower Grand Canyon, yet it is likely more widespread.

Native people use this and other members of the genus *Artemisia* ceremonially and medicinally. The leaves and stems are boiled into a tea, burned as an incense for purification, or crushed and used as a poultice.

BACCHARIS *Baccharis* species
SUNFLOWER FAMILY (Asteraceae/Compositae)

Baccharis are common shrubs throughout the river corridor and in many side canyons, where their spreading root systems effectively stabilize the soil. They are distributed from California east to Texas and south into mainland Mexico, Baja California, and South America. Baccharis can be separated into two groups: the broad-leaved, tall shrubs called *seep willows* and the nearly leafless shrubs known as *desert brooms*. In the Grand Canyon, they are widespread along the river and in side canyons throughout the Inner Canyon.

Baccharis species are widely used by many indigenous people. The branches are used for basketry, arrows, brooms, roof and building material, and fuel wood. The leaves are medicinal for bruises and wounds. Young shoots may be eaten when nothing else is available. Ground and boiled roots were used for flavoring wine, which may have lead to the naming of the genus, after *Bacchus*, who was the Greek god of wine.

Louisiana Wormwood *Artemisia ludoviciana*
—Photograph © 2006 by David Edwards

Desert Broom *Baccharis sarothroides*
—Photograph © 2006 by Lisa A. Hahn

Desert Broom *Baccharis sarothroides* —Photograph © 2006 by Celia Southwick

Emory's Seep-Willow *Baccharis emoryi*

Plants: Openly branched shrub, 1 to 5 meters tall. **Leaves:** Alternate, linear to oblanceolate, with three to five large blunt teeth at the tip, blue green, three-veined, to 7 centimeters long. **Flowers:** In diffuse, elongated panicles; disk flowers white, 4 to 8 millimeters long; ray flowers lacking. **Fruit:** Achene with pappus bristles. **Flowering Season:** April to September. **Elevation:** 1,200 to 3,700 feet.

Baccharis, Seep-Willow *Baccharis salicifolia* (*B. glutinosa*)

Plants: Openly branched shrub, 1 to 5 meters tall. **Leaves:** Alternate, linear to lanceolate, entire to finely toothed, shiny green, one- to three-veined, to 15 centimeters long. **Flowers:** In compact, rounded panicles; disk flowers white, 3 to 6 millimeters long; ray flowers lacking. **Fruit:** Achene with pappus bristles. **Flowering Season:** March to December. **Elevation:** 1,200 to 3,500 feet.

Both species of seep-willow are showy in late fall when the white pappus of the fruits is prominent. Although sometimes confused with willow (*Salix* species), the green longitudinal stripes along the stems and the plumose pappus distinguish seep-willows. Emory's seep-willow is most easily confused with baccharis, especially where their distributions overlap. These two species may hybridize, making them even more difficult to tell apart. The leaves of baccharis are usually shiny green with a sometimes sticky or resinous surface whereas the leaves of Emory's seep-willow are slightly bluish green. Baccharis also has a much more astringent, persistent taste and a shorter plumose pappus than Emory's seep-willow.

Emory's seep-willow was named for Lieutenant William Hemsley Emory (1811–1887) who followed the Gila River from its source in New Mexico to its confluence with the Colorado River during the early 1800s. Emory faithfully kept a journal and collected plants, which the Smithsonian now houses.

Desert Broom *Baccharis sarothroides*

Plants: Erect, broomlike shrub, 1 to 4 meters tall. **Leaves:** Linear to narrowly oblanceolate, reduced to small bracts on upper stems, one-veined, to 3.5 centimeters long. **Flowers:** In panicles; disk flowers cream, 3 to 8 millimeters long; ray flowers lacking. **Fruit:** Achene with pappus bristles. **Flowering Season:** October to March. **Elevation:** 1,200 to 2,100 feet.

Desert Baccharis *Baccharis sergiloides*

Plants: Compact, rounded shrub, to 2 meters tall. **Leaves:** Oblanceolate to obovate, entire or few toothed, reduced to small bracts on upper stems, one-veined, to 3.5 centimeters long. **Flowers:** In dense panicles; disk flowers cream, 2 to 5 millimeters long; ray flowers lacking. **Fruit:** Achene with pappus bristles. **Flowering Season:** August to October. **Elevation:** 1,200 to 5,200 feet.

Desert broom and desert baccharis differ from the seep-willows by their dense branching and reduced leaves, and also their preference for drier slopes. Desert broom has narrower leaves and grows twice as tall as desert baccharis, with branches reaching upward. Desert baccharis forms mounded shrubs with intricate branches and has more spoon-shaped leaves. Unlike its more widespread baccharis relatives, desert broom first appears around River Mile 125 and becomes especially abundant below National Canyon, where it stands tall behind the trees and shrubs along the river's edge.

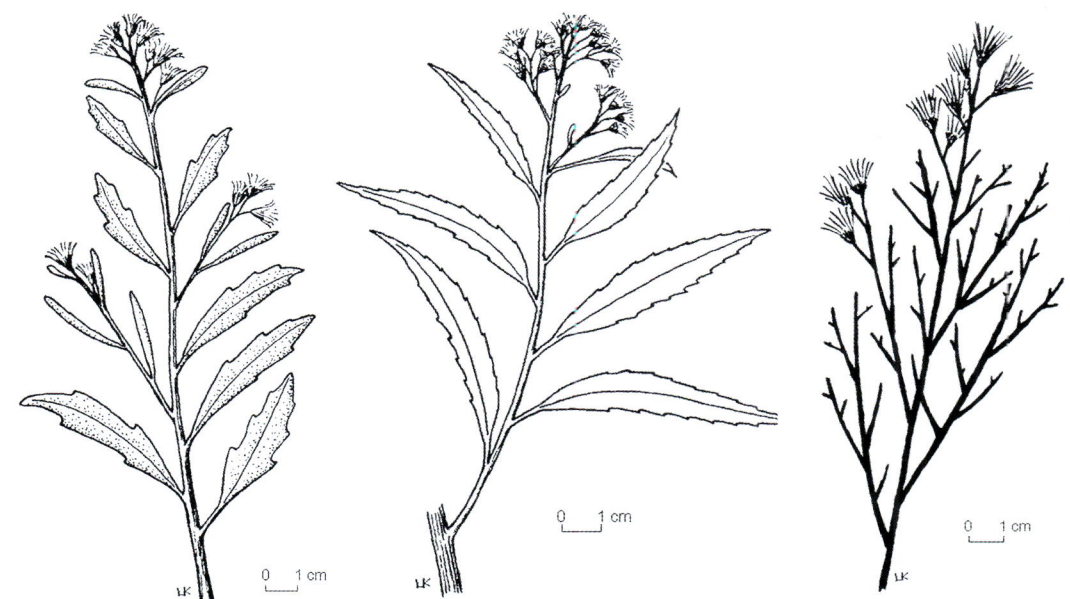

Emory's Seep-Willow
Baccharis emoryi
—Illustration © 2006 by Lisa Kearsley

Baccharis
Baccharis salicifolia
—Illustration © 2006 by Lisa Kearsley

Desert Broom
Baccharis sarothroides
—Illustration © 2006 by Lisa Kearsley

Desert Baccharis
Baccharis sergiloides
—Illustration © 2006 by Lisa Kearsley

Baccharis, Seep-Willow *Baccharis salicifolia*
—Photograph © 2006 by Kate Watters

Desert Marigold
Baileya multiradiata
SUNFLOWER FAMILY (Asteraceae/Compositae)

Plants: Annual or short-lived perennial, 5 to 60 centimeters tall. **Leaves:** Alternate, pinnate, grayish green, 2 to 11 centimeters long. **Flowers:** In solitary or few-flowered cymes, to 30 centimeters tall; disk flowers yellow; ray flowers yellow, three-lobed. **Fruit:** Achene. **Flowering Season:** April to August. **Elevation:** 1,200 to 4,200 feet.

Desert marigold is one of the most widespread and easily recognized wildflowers of the desert Southwest, with huge, bright yellow flowering heads. Long, white, woolly hairs cover the entire plant. This common adaptation helps desert plants reflect light and minimize moisture loss, thereby protecting the plant from extreme heat and drought. Desert marigold responds quickly to rain and blooms throughout the year, although most profusely in the spring and summer.

Desert marigold grows from California east to Texas and south to Mexico, often on roadsides and in disturbed areas. In the Grand Canyon, it grows the length of the river, but is most common in the lower, warmer reaches of the Inner Canyon.

One species of schinia moth (*Schinia miniana*) lays its eggs on the flowers. When the caterpillar is ready to form a cocoon, it wraps the flower petals around it for protection. The yellow moths blend in well with the flowers, which probably serves to protect them from predators. Desert marigold is a popular xeriscape garden plant because it is drought tolerant and easily grown from seed.

The genus *Baileya* honors Jacob Whitman Bailey (1811–1914), who was an early pioneer of the microscope and professor of chemistry at West Point Military Academy.

Chuckwalla's Delight
Bebbia juncea var. *aspera*
SUNFLOWER FAMILY (Asteraceae/Compositae)

Plants: Much-branched, rounded shrub, to 1.5 meters tall. **Leaves:** Opposite, linear, dark green, rough hairy, to 7 centimeters long. **Flowers:** In solitary or clustered cymes, to 7 centimeters tall; disk flowers yellow; ray flowers lacking. **Fruit:** Achene. **Flowering Season:** March to November. **Elevation:** 1,200 to 3,100 feet.

Chuckwalla's delight, a sparse, straggly shrub with sweet-smelling flowers, blooms all year long. An intricate array of stems spreads from a woody base. The minute leaves are short-lived and shed early to minimize water loss; however, the smooth, green stems facilitate year-round photosynthesis.

You might confuse chuckwalla's delight with spiny aster (*Chloracantha spinosa*) because both have inconspicuous, reduced leaves and spindly, green stems. Spiny aster has white ray and yellow disk flowers and grows in clumps of narrowly upright spiny stems. The rounder, more densely branched chuckwalla's delight grows on drier slopes and bears only yellow disk flowers.

Chuckwalla's delight ranges from California to Texas and Mexico and is the only species of *Bebbia* in Arizona. In the Grand Canyon, it grows on beaches and rocky slopes above the river's edge from Badger Canyon to Lake Mead.

The chuckwalla (*Sauromalus obesus*), a large herbivorous lizard, savors the flower heads, which provide a high-energy food. According to *Natural History of the Sonoran Desert*, naturalist Edmund Jaegar (1887–1983) named the shrub when he observed "chuckwallas feeding greedily on the flowers." The yellow disk flowers host numerous species of native bees and butterflies, including the Yuma skipper (*Ochlodes yuma*) and western skipperling (*Oarisma garita*). The Spanish name for the plant, *chuparosa*, means "sucked for nectar."

Desert Marigold *Baileya multiradiata* —Photograph © 2006 by Kate Watters
Inset: Schinia moth feeding on Desert Marigold —Photograph © 2006 by John Running

Chuckwalla's Delight *Bebbia juncea* var. *aspera*
—Photograph © 2006 by Tyler Williams

Chuckwalla's Delight *Bebbia juncea* var. *aspera*
—Illustration © 2006 by Mar-Elise Hill

BRICKELLBUSH
Brickellia species

SUNFLOWER FAMILY (Asteraceae/Compositae)

The most definite way to identify the genus *Brickellia* is by the bitter taste of the leaves, which resembles that of a grapefruit rind. Dramatically different leaf shapes and margins distinguish the three most common species of brickellbush in the Grand Canyon. The genus *Brickellia* honors John Brickell (1749–1809), an Irish physician and botanist from Savannah, Georgia.

Long-Leaf Brickellbush
Brickellia longifolia

Plants: Upright shrub, to 1.5 meters tall. **Leaves:** Alternate, entire, linear, inconspicuously veined, 1 to 14 centimeters long. **Flowers:** In racemes; disk flowers creamy white to green; ray flowers lacking. **Fruit:** Achene with pappus of capillary bristles. **Flowering Season:** August to October. **Elevation:** 1,900 to 6,900 feet.

As fall arrives, fragrant, creamy white blossoms cloak the extremely common long-leaf brickellbush. This resilient shrub with bright green, varnished foliage is often the first to regenerate after a flash flood, due largely to its deep root system and ability to sprout from the root crown. Long-leaf brickellbush has long, narrow leaves that bear no teeth and have inconspicuous veins. It occurs in Colorado, Utah, Arizona, Nevada, and California. In the Grand Canyon, it grows on rocky, talus slopes, and cliffs and is one of the most common plants along dry drainages and river tributaries.

Spiny Brickellbush
Brickellia atractyloides

Plants: Compact shrub, 20 to 30 centimeters tall. **Leaves:** Alternate, triangular to ovate, sharp-pointed margins, spine-tipped, prominently veined, 5 to 30 millimeters long. **Flowers:** In solitary, terminal heads; disk flowers creamy white to pink; ray flowers lacking. **Fruit:** Achene with pappus of capillary bristles. **Flowering Season:** April to May. **Elevation:** 1,200 to 5,500 feet.

Although striking when in full bloom, you are most likely to notice spiny brickellbush when it has past its prime. The flower bracts and leathery leaves turn chalky gray with age, remaining on the plant throughout the year. Spiny brickellbush has thick, leathery leaves with distinct veins, spiny tips, and sharp-pointed margins. Spiny brickellbush ranges from southern California and Baja California to Nevada, Utah, and Arizona. In the Grand Canyon, plants are common throughout the Inner Canyon, where they favor schist walls and talus slopes.

California Brickellbush
Brickellia californica

Plants: Rounded shrub, to 2 meters tall. **Leaves:** Alternate, triangular to ovate, toothed margins, distinct petioles, 1 to 5 centimeters long. **Flowers:** In panicles; disk flowers greenish white to pale yellow; ray flowers lacking. **Fruit:** Achene with pappus of capillary bristles. **Flowering Season:** August to September. **Elevation:** 1,700 to 7,800 feet.

California brickellbush, a staple plant of arid landscapes, has widely triangular to ovate, toothed leaves that lack sharp tips and pointy teeth. Many cultures throughout western North America use this shrub to control blood sugar levels. It occurs from Oregon and Idaho and south to Mexico. In the Grand Canyon, plants are well distributed but never abundant, from the vicinity of Nankoweap Creek to Lava Falls.

Another less common species, **Coulter's brickellbush (*Brickellia coulteri*),** grows in similar habitats in the Grand Canyon. It has more distinctly pointed triangular leaves that are thinner and less stiff than those of California bricklellbush.

Long-Leaf Brickellbush *Brickellia longifolia*
—Illustration © 2006 by Bronze Black

Long-Leaf Brickellbush *Brickellia longifolia*
—Photograph © 2006 by Celia Southwick

Spiny Brickellbush *Brickellia atractyloides*
—Photograph © 2006 by Arizona Raft Adventures

Spiny Brickellbush *Brickellia atractyloides*
—Photograph © 2006 by Kate Watters

Desert Pincushion *Chaenactis stevioides*
SUNFLOWER FAMILY (Asteraceae/Compositae)

Plants: Annual herb, to 45 centimeters tall. **Leaves:** Alternate, linear, pinnately lobed, to 11 centimeters long. **Flowers:** Solitary or in panicles; disk flowers white; ray flowers lacking. **Fruit:** Achene with scaly pappus. **Flowering Season:** March to May. **Elevation:** 2,600 to 4,000 feet.

Among Grand Canyon's early spring bloomers is desert pincushion, a woolly annual. The sparsely leaved, lanky stems display dense clusters of tiny white, disk flowers. The outer disk flowers appear raylike and are actually larger than those found in the head's center. Their stamens prominently protrude from the flower centers.

Desert pincushion occurs from Oregon and Idaho south to New Mexico and Arizona, and into Mexico. In the Grand Canyon, it grows on sandy beaches and desert slopes. It is known from Lees Ferry to Hance Rapid but likely occurs farther downriver.

Spiny Aster *Chloracantha spinosa*
SUNFLOWER FAMILY (Asteraceae/Compositae) (*Aster spinosus*)

Plants: Erect perennial herb, to 1.5 meters tall. **Leaves:** Alternate, linear to spatulate, spiny, 1 to 5 centimeters long. **Flowers:** Solitary or in loosely clustered panicles; disk flowers yellow; ray flowers white. **Fruit:** Achene with barbed bristly pappus. **Flowering Season:** May to December. **Elevation:** 1,200 to 3,500 feet.

Spiny aster, a broomlike herb, forms dense thickets on the beaches and cobblebars of the Colorado River. Initially colonizing from tiny, wind-borne seeds, it then spreads aggressively through deep rhizomes. The green, spiny stems allow the plant to photosynthesize despite its reduced leaves. The dainty, daisylike flowers attract bees and butterflies.

Spiny aster is easily confused with Chuckwalla's delight (*Bebbia juncea* var. *aspera*), especially when the plants lack flowers, because both are a maze of almost leafless, green stems. Spiny aster grows in bright green diffuse patches near the water's edge and has spine-tipped leaves.

Spiny aster has a broad geographical distribution, tolerating tropical to warm temperate climates in California, Nevada, and Utah, and south to Central America. In the Grand Canyon, it occurs on beaches, shoreline marshes, and in side canyons the length of the river.

The genus name, *Chlorocantha*, is from the Greek *chloros*, meaning "green," and *akantha*, meaning "thorn" or "prickle."

Desert Pincushion *Chaenactis stevioides* —Photograph
© 2006 by Daniela Roth/Navajo Natural Heritage Program

Desert Pincushion *Chaenactis stevioides*
—Photograph © 2006 by David Inouye

Spiny Aster *Chloracantha spinosa*
—Photograph © 2006 by Kate Watters

Spiny Aster *Chloracantha spinosa*
—Illustration © 2006 by Mar-Elise Hill

New Mexico Thistle *Cirsium neomexicanum*
SUNFLOWER FAMILY (Asteraceae/Compositae)

Plants: Stout, spiny-stemmed, biennial herb, to 2 meters tall. **Leaves:** Alternate, lanceolate to elliptic, gray green, deeply lobed, spine-tipped, densely hairy, to 40 centimeters long. **Flowers:** In cymelike panicles, with spiny bracts; disk flowers white to lavender; ray flowers lacking. **Fruit:** Achene with bristly pappus. **Flowering Season:** April to May. **Elevation:** 2,500 to 6,500 feet.

New Mexico thistle is one of ten native thistles in the Grand Canyon. Numerous other thistle species are highly invasive exotics. The exotic species have spiny wings along the stem that continue well below the leaf base, a characteristic that native species typically do not share.

New Mexico thistle thrives at many elevations throughout the desert Southwest from Colorado west to California. In the Grand Canyon, it flourishes in washes and dry rocky slopes from desertscrub to piñon-juniper woodlands. Look for it at Buck Farm, Clear Creek, Phantom Ranch, and Havasu Canyon.

Three other notable native *Cirsium* species grow in the Inner Canyon. **Rydberg thistle (*C. rydbergii*)** is hard to misidentify because its robust leathery basal leaves grow to 90 centimeters long and its flowering stalk reaches heights of nearly 2 meters. This pink-flowered thistle grows in hanging gardens and seeps in Buck Farm and Saddle Canyons. A large, spiny thistle, which is abundant at Three Springs, bears a striking resemblance to Rydberg thistle and is likely **Mojave thistle (*C. mohavense*)**. **Cainville thistle (*C. arizonicum* var. *bipinnatifita*)** is shorter than Rydberg and New Mexico thistles and is known from only a few locations, such as Thunder River and springs on both rims.

The treacherous spines of thistles do not deter pollinators from visiting the flowers, which provide nectar for many species, including hummingbirds, beetles, bumblebees, butterflies, and hawk moths. Birds eat the seeds and use the down to build their nests. The genus name, *Cirsium*, derives from the Greek *cirsos*, meaning "swollen vein," referring to the use of thistle plants as a remedy for varicose veins.

New Mexico Thistle *Cirsium neomexicanum*
—Photograph © 2006 by Kyle George

Rydberg Thistle *Cirsium rydbergii*
—Photograph © 2006 by David Edwards

New Mexico Thistle *Cirsium neomexicanum* —Photograph © 2006 by Kathy Darrow

Horseweed
Conyza canadensis
SUNFLOWER FAMILY (Asteraceae/Compositae)

Plants: Exotic, erect annual herb, to 2 meters tall. **Leaves:** Alternate, lanceolate to oblanceolate, lower leaves toothed, with stiff hairs, 1 to 10 centimeters long. **Flowers:** In tall panicles, with numerous heads; disk flowers yellow; ray flowers white or purple;. **Fruit:** Achene with white bristles. **Flowering Season:** April to November. **Elevation:** 1,200 to 8,500 feet.

When in flower or seed, horseweed is easy to recognize because of its tall, dense stalks bearing many flower heads. It is more difficult to identify early in its life cycle when only basal rosettes cover the ground. The slender bristles at the top of horseweed seeds allow for ready dispersal by wind, with a single plant capable of producing up to 60,000 seeds. This reproductive success has helped horseweed extend its range into new territory, such as the Grand Canyon.

Horseweed is native to eastern North America. In the Grand Canyon, it is now common in side canyons like Deer Creek and along the river corridor from Lees Ferry to Lake Mead. It thrives on both rims, in woodlands and forests, and also in disturbed areas like drainages, debris fans, and grasslands, especially following floods and fires.

The leaves and flowers contain terpenes, aromatic compounds that may irritate the skin and respiratory systems of humans and animals. Horseweed may be poisonous if consumed. When dried, the aerial portions of the plant can be used as a skin astringent.

Desert Dicoria, Desert Twin Bugs
Dicoria canescens
SUNFLOWER FAMILY (Asteraceae/Compositae)

Plants: Annual herb, to 1 meter tall. **Leaves:** Lanceolate to triangular-ovate, gray green, toothed, covered with hair, 1 to 6 centimeters long. **Flowers:** In panicles, heads enclosed from below by bracts; disk flowers cream to green or purple; ray flowers lacking. **Fruit:** Winged achene. **Flowering Season:** August to March. **Elevation:** 1,200 to 2,900 feet.

Look for desert dicoria, a fragrant, sparsely leaved, many-branched annual, growing on sandy beaches amidst sand-verbena (*Abronia elliptica*) and pale evening-primrose (*Oenothera pallida*). The leaves of the young plants often drop before the flowers bloom, lending to its spare appearance. When fully mature, it resembles tumbleweed as it rolls and tumbles across the sand.

Desert dicoria inhabits warmer areas from California, east to Colorado and New Mexico, and south to Mexico. In the Grand Canyon, this plant is represented by two subspecies (ssp. *canescens* and ssp. *brandegeei*) and is commonly found from River Mile 38 to 232.

Archaeological evidence suggests that prehistoric people ate the large seeds of desert dicoria. Because the seeds are available in the winter, they provided an important food, supplementing stored food reserves. The genus name, *Dicoria*, comes from the Greek *dis*, meaning "twice," and *koris*, meaning "bug," referring to the fruits, which occur in pairs and look like flat-bodied bugs.

Horseweed *Conyza canadensis*
—Photograph © 2006 by David Edwards

Desert Dicoria *Dicoria canescens* —Photograph © 2006 by David Edwards

# Brittlebush	*Encelia farinosa*
SUNFLOWER FAMILY (Asteraceae/Compositae)

Plants: Rounded shrub, to 1.6 meters tall. **Leaves:** Alternate, triangular to ovate, silvery gray, silky hairy, 2 to 10 centimeters long. **Flowers:** Solitary or in cymelike panicles, heads on naked stalks; disk and ray flowers yellow. **Fruit:** Hairy achene. **Flowering Season:** February to June. **Elevation:** 1,200 to 4,100 feet.

Brittlebush heralds the onset of spring in the desert, particularly during years of adequate winter moisture, when its flowers transform hillsides into a riotous yellow. As the season progresses, this deciduous shrub disperses its seeds, drops its silvery gray leaves, and slips into dormancy where it awaits the next rain. At this stage, you can identify brittlebush by its intricately branched architecture and bleached white stems.

Resin brittlebush (*Encelia resinifera*) has smaller, resin-coated, dark green leaves. Both have yellow flower heads, but those of resin brittlebush are typically smaller and solitary on the stem, unlike the clustered flower heads of brittlebush. Some plants exhibit characteristics of both species and may be hybrids.

Brittlebush grows in Arizona, Utah, Nevada, California, and northern Mexico. In the Grand Canyon, it occurs mainly below the Tonto Plateau on rocky desert slopes and washes. Brittlebush is abundant from River Mile 40 downstream to Grand Wash Cliffs. Like other Sonoran Desert plants, this shrub uses the Colorado River as a corridor for upstream expansion, reaching cooler, frost-prone areas at the edge of its range.

Native bees use the stems as a nighttime roost. The stems exude a fragrant resin useful for waterproofing pots and wood, and as incense in churches, leading to its other common name, *incienso*.

# Resin Brittlebush	*Encelia resinifera*
SUNFLOWER FAMILY (Asteraceae/Compositae)

Plants: Small to medium shrub, to 1.6 meters tall. **Leaves:** Alternate, ovate to narrowly oblong, mostly entire, to 2.5 centimeters long. **Flowers:** In solitary heads on leafless stalks; ray and disk flowers yellow. **Fruit:** Achene. **Flowering Season:** April to November. **Elevation:** 1,700 to 5,000 feet.

The dark green, shiny leaves of resin brittlebush are coated with resin and emit an aroma when touched. These shrubs are most robust where soil moisture is abundant, such as in desert washes. The showy yellow flowers attract butterflies, solitary bees, and occasional honeybees, which transport pollen between plants.

Resin brittlebush is the foremost *Encelia* of the Colorado Plateau. A Mojave Desert native, it inhabits the warm areas of Arizona, Utah, Nevada, and California. In the Grand Canyon, it is common on rocky slopes and in washes, from the Supai Formation down to the Tapeats sandstone, where it is replaced by its more drought-tolerant relative, brittlebush (*Encelia farinosa*). A less common, narrow-leaved, narrow-rayed subspecies, **E. resinifera ssp. tenuifolia**, is endemic to the Grand Canyon.

The genus is named after Christopher Entzelt (1517–1583), a Lutheran clergy who latinized his name to Encelium and published books on medicinal uses of animals and plants.

Brittlebush *Encelia farinosa* —Photograph © 2006 by Roger Dale

Brittlebush *Encelia farinosa*
—Photograph © 2006 by Raechel Running RMRfotoarts.com

Resin Brittlebush *Encelia resinifera*
—Photograph © 2006 by David Edwards

Fleabane Daisy
Erigeron lobatus

SUNFLOWER FAMILY (Asteraceae/Compositae)

Plants: Annual, biennial, or perennial herb, 20 to 40 centimeters tall. **Leaves:** Alternate, obovate, four to eight lobed, with sparsely spreading hairs, 3 to 10 centimeters long. **Flowers:** Solitary or in panicles; disk flowers yellow; ray flowers white, lavender, or blue. **Fruit:** Achene with fluffy white pappus. **Flowering Season:** March to September. **Elevation:** 1,500 to 3,600 feet.

Fleabane daisy has delicate, dark green foliage and lavender blue ray flowers. The best way to distinguish this genus from others in the sunflower family with whitish lavender flowers is to note the number of small, narrow ray flowers. If there are too many to count, it is probably a fleabane daisy. Our species has broad, glandular leaves that are divided into relatively wide lobes, a pattern most prominent on the lower leaves. This differentiates it from other nonglandular, hairy, narrow-leaved fleabane daisies.

Fleabane daisy was once thought to be endemic to Arizona but has now been found in Nevada, Utah, and California. One of fourteen *Erigeron* species in the Grand Canyon, it is common throughout the Inner Canyon and in side canyons, often on sandy, moist soil on beaches and at cliff bases.

The genus name, *Erigeron*, is from the Greek words *eri*, meaning "early," and *geron*, meaning "old man," referring to the fluffy nature of the early spring seed heads.

Gray Aster
Eurybia glauca
(*Aster glaucodes*)

SUNFLOWER FAMILY (Asteraceae/Compositae)

Plants: Perennial herb, to 50 centimeters tall. **Leaves:** Alternate, linear to lanceolate, pale green, glaucous, 3 to 7 centimeters long. **Flowers:** In panicles; disk flowers yellow; ray flowers white to violet. **Fruit:** Achene with bristly pappus. **Flowering Season:** September to November. **Elevation:** 1,800 to 3,600 feet.

The showy, violet-tinged, white flowers bring large, dense patches of gray aster to life. Its firm, wide leaves appear grayish green because of their powdery coating. Gray aster occurs from Montana and Idaho south to Arizona and New Mexico. In the Grand Canyon, it is common in side canyons and alcoves at scattered locations the length of the river.

The genus name, *Eurybia*, is from the Greek word *Eurybies*, meaning "far" and "wide spreading," referring to its spreading habit.

Fleabane Daisy *Erigeron lobatus* —Photograph © 2006 by David Inouye
Inset —Photograph © 2006 by David Inouye

Gray Aster *Eurybia glauca* —Photograph by Lori Makarick/National Park Service

Western Goldenrod   *Euthamia occidentalis*
SUNFLOWER FAMILY (Asteraceae/Compositae)   (*Solidago occidentalis*)

Plants: Perennial herb, to 2 meters tall. **Leaves:** Alternate, narrowly lanceolate, sessile, 2 to 10 centimeters long. **Flowers:** In cymelike panicles; disk and ray flowers yellow. **Fruit:** Achene with white, bristly pappus. **Flowering Season:** June to October. **Elevation:** 1,700 to 3,100 feet.

Western goldenrod flourishes on riverbanks, beaches, and marshes with the help of its wide-spreading rhizomes. In the fall, when the bright yellow blossoms emerge, it is the most prominent flowering plant along the shoreline. Its dense, dark green, aromatic foliage helps differentiate it from other riverside species in the sunflower family. Look for its erect, open architecture and clusters of small, yellow flowers.

Western goldenrod occurs from Baja California into Canada, and east through the Rocky Mountain States. In the Grand Canyon, it is widely distributed in the riparian zone the length of the river.

The genus name, *Euthamia*, is Greek for "well-crowded," which refers to the densely packed flowering stalks.

Grand Canyon Flaveria   *Flaveria macdougallii*
SUNFLOWER FAMILY (Asteraceae/Compositae)

Plants: Perennial herb, to 1 meter tall. **Leaves:** Opposite, linear, succulent, 5 to 14 centimeters long. **Flowers:** In cymes; disk flowers yellow; ray flowers lacking. **Fruit:** Achene. **Flowering Season:** September to November. **Elevation:** 1,670 to 1,800 feet.

Grand Canyon flaveria was first described in 1975 by M. E. Thereaux. This species clings to the walls of limestone seeps, bursting with golden blossoms late in the fall when it becomes a conspicuous element of hanging gardens. It often forms dense patches because it has woody rhizomes, and fallen stems root at the nodes. Plants in the *Flaveria* genus have up to four different chemical pathways for completing photosynthesis.

Grand Canyon flaveria is endemic to the Grand Canyon. It is restricted to saline springs, seeps, and moist open slopes on Muav limestone benches above the river and in side canyons, from just upstream of Matkatamiba Creek downstream to Cove Canyon.

The genus name, *Flaveria*, is from the Latin *flavus* for "yellow," in reference to the yellow flowers. Although spelling his name incorrectly, the epithet, *macdougallii*, honors Walter B. McDougall (1883–1980), a former curator of the Museum of Northern Arizona, who wrote several floras of Arizona.

Western Goldenrod *Euthamia occidentalis* —Photograph © 2006 by Celia Southwick

Grand Canyon Flaveria *Flaveria macdougalii* —Photograph © 2006 by Burns Jansen

Broom Snakeweed *Gutierrezia sarothrae*
SUNFLOWER FAMILY (Asteraceae/Compositae)

Plants: Rounded shrub, to 90 centimeters tall. **Leaves:** Alternate, linear to lanceolate, 2 to 7 centimeters long. **Flowers:** Solitary or in panicles; disk and ray flowers yellow. **Fruit:** Hairy achene with scaly pappus. **Flowering Season:** May to November. **Elevation:** 1,200 to 8,000 feet.

In the western United States, snakeweeds are indicators of disturbance or overgrazing on range and wildlands. These toxic and resinous plants are often the lone survivors among their more palatable neighbors.

Most commonly found in the western United States and Canada, broom snakeweed also grows in New York and Minnesota. Throughout the Inner Canyon, it inhabits talus slopes, rocky areas, and dry soils in woodlands and shrublands.

Broom snakeweed's range overlaps with **threadleaf snakeweed (***Gutierrezia microcephala***)** throughout most of the Grand Canyon. They are nearly impossible to distinguish without flowers, but threadleaf snakeweed tends to be more abundant in the western canyon. When in flower, note that threadleaf snakeweed has one to two ray flowers per head compared to the three to seven ray flowers per head of broom snakeweed.

Broom snakeweed can be confused with small forms of **rabbitbrush (***Ericameria nauseosa***)**, but snakeweed leaves are greener and stinkier. Rabbitbrush also has much larger flowering heads than snakeweed and is much less common at the lower elevations in the Inner Canyon.

Snakeweeds have many medicinal uses, especially to relieve muscle soreness. The flowers produce a yellow dye. The genus name derives from a Spanish noble family Gutierrez. The specific epithet, *sarothrae*, is from the Greek *sarum*, meaning "broom," which is appropriate as the branches make good brooms and whisks.

Willow Glowweed *Hesperodoria salicina*
SUNFLOWER FAMILY (Asteraceae/Compositae) (*Haplopappus salicinus*)

Plants: Perennial shrub, to 50 centimeters tall. **Leaves:** Alternate, linear-lanceolate, 1 to 7 centimeters long. **Flowers:** Solitary or in panicles; disk flowers yellow; ray flowers lacking. **Fruit:** Achene with bristly pappus. **Flowering Season:** September to November. **Elevation:** 1,200 to 4,700 feet.

Willow glowweed is only known to grow within the boundaries of the Grand Canyon. Sunken glands dot the wispy, lime green foliage that glows in the afternoon sunlight. Willow glowweed resembles **Grand Canyon goldenbush (***Hesperodoria scopulorum* **var.** *scopulorum***)**, which lacks resin-producing glands and is less common in the Grand Canyon. These are the only two species in this genus.

Willow glowweed is not very conspicuous, but a discerning eye might find it growing on shaded limestone walls and boulders near seeps and along trails the length of the river. Look for it at Saddle Canyon and Carbon Canyon (River Mile 64.5).

The origin of the genus name, *Hesperodoria*, is uncertain. It may be from the Greek word *Hesperus*, which may refer to the belief that the planet Venus was originally thought to be two objects, one of which was called Hesperus. Or the genus may be from the Greek myth of the Hesperides, the naughty nymphs of the evening, who along with a dragon protected Zeus's and Hera's golden apples until Heracles tricked Atlas into stealing them for him.

Broom Snakeweed *Gutierrezia sarothrae* —Photograph © 2006 by David Edwards
Inset —Photograph © 2006 by Max Licher

Rabbitbrush *Ericameria nauseosa*
—Photograph © 2006 by Glenn Rink

Willow Glowweed *Hesperodoria salicina*
—Photograph © 2006 by Kate Watters

Goldenbush, Jimmyweed *Isocoma acradenia*
SUNFLOWER FAMILY (Asteraceae/Compositae)

Plants: Erect shrub, to 1.3 meters tall. **Leaves:** Alternate, linear to spatulate, toothed, to 6 centimeters long. **Flowers:** In cymes; disk flowers yellow; ray flowers lacking. **Fruit:** Achene. **Flowering Season:** July to November. **Elevation:** 1,200 to 3,300 feet.

Not until its yellow flowers blossom in late summer does it become apparent that goldenbush is one of the most abundant plants in the Inner Canyon. During other times of the year, its straight, gray branches and aromatic dark green leaves blend in with other desert shrubs.

Goldenbush grows in alkaline soils in the Sonoran and Mojave Deserts in California, Nevada, Utah, Arizona, and Baja California. In the Grand Canyon, it grows in patches on slopes and flats, in side canyons, and on active travertine slopes from around Saddle Canyon downstream to Grand Wash Cliffs.

Goldenbush leaves are used as a poultice for sores and muscular pain and are chewed for coughs. Some native people employed the plant for building fences and as kindling. The epithet, *acradenia*, is Greek for "point-glanded," referring to the bracts that bear a large gland at the tip.

Spiny Goldenweed *Machaeranthera pinnatifida*
SUNFLOWER FAMILY (Asteraceae/Compositae) (*Haplopappus spinulosus*, in part)

Plants: Perennial herb, 12 to 60 centimeters tall. **Leaves:** Alternate, narrowly linear, entire to toothed, 1 to 6 centimeters long. **Flowers:** Solitary or in cymes, disk and ray flowers yellow. **Fruits:** Hairy achene with tan, bristly pappus. **Flowering Season:** January to October. **Elevation:** 1,200 to 6,100 feet.

Spiny goldenweed, also called *lacy tansy aster*, qualifies as one of the Grand Canyon's many DYCs or "Damned Yellow Composites," a term used by naturalists frustrated by the dozens of yellow-flowered members of the sunflower family. However, most are really not so hard to identify if you look for a few key characteristics. To identify spiny goldenweed, look for bristly tips on the small leaves and bracts enclosing the flower head. It also has larger flower heads and lobed leaves with spines on the margin compared to the smaller heads and needle-like leaves of fetid marigold (*Thymophylla pentachaeta* var. *pentachaeta*). The annual desert-dandelion (*Malacothrix glabrata*) blooms much earlier and has a distinctive, milky sap. Other similar *Machaeranthera* species grow primarily at higher elevations and have ray flowers that vary from yellow to purple.

The six varieties and two subspecies of spiny goldenweed occur throughout the West from central Canada to Mexico except in Washington and Idaho, and at scattered locations east of the Mississippi River. In the Grand Canyon, plants have a wide elevational range and are common in dry gravelly soils, crevices, canyon bottoms, and on slopes from Lees Ferry to Grand Wash Cliffs.

Goldenbush *Isocoma acradenia* —Photograph © 2006 by Max Licher

Goldenbush *Isocoma acradenia*
—Photograph © 2006 by Max Licher

Spiny Goldenweed *Machaeranthera pinnatifida*
—Photograph © 2006 by Kristin Huisinga

Desert-Dandelion
SUNFLOWER FAMILY (Asteraceae/Compositae)

Malacothrix glabrata
(*M. californica* var. *glabrata*)

Plants: Winter annual or biennial, 5 to 60 centimeters tall. **Leaves:** Basal, linear to oblong, with filiform lobes, glandular to hairy, to 15 centimeters long. **Flowers:** Solitary or few, heads on long stalks; disk flowers lacking; ray flowers yellow, with five teeth. **Fruit:** Achene with white pappus. **Flowering Season:** March to June. **Elevation:** 1,200 to 4,000 feet.

After a wet winter, look for large displays of blooming desert-dandelion, especially on the Dox Formation between Tanner and Unkar. The sunny flower head has bright red centers until all its yellow ray flowers unfurl. The milky sap is one of the best clues to its identity, a characteristic it shares with its weedy cousin the **common dandelion (***Taraxacum officinale***)**. Common dandelion has broad, basal leaves and is much shorter.

Desert-dandelion ranges from Oregon and Idaho south to Mexico. In the Grand Canyon, you will find it in coarse soils from Lees Ferry throughout the Inner Canyon downstream to Rampart Cave at River Mile 275.

A less common relative, **yellow saucers (***Malacothrix sonchoides***)**, typically grows at higher elevations, although it has been collected along the river. Its leaves are wider and coarsely toothed unlike the threadlike lobes of desert-dandelion leaves.

The common name comes from the French word *dent-de-lion*, meaning "lion's teeth," describing the toothed ray flowers.

Emory Rock Daisy
SUNFLOWER FAMILY (Asteraceae/Compositae)

Perityle emoryi

Plants: Robust, much-branched, winter annual herb, 10 to 60 centimeters tall. **Leaves:** Alternate, triangular to ovate, deeply toothed and lobed, 2 to 10 centimeters long. **Flowers:** Solitary or in loosely clustered cymes; disk flowers yellow; ray flowers white. **Fruit:** Hairy achene. **Flowering Season:** April to May. **Elevation:** 1,500 to 3,000 feet.

Emory rock daisy teems with cheery, compact daisylike flowers. The broad, vivid green leaves are covered in glandular hairs and resemble a geranium. At first glance, this plant is similar to spiny aster (*Chlorocantha spinosa*), which also has daisylike flowers, yet lacks the glandular, geraniumlike leaves of Emory rock daisy.

Emory rock daisy occurs from California east to Arizona and south to Mexico and South America. In the Grand Canyon, it grows on rocky slopes and cliffs along the river from the confluence of the Little Colorado River to Cardenas Creek (near River Mile 71), and at scattered locations downstream to Grand Wash Cliffs.

Grand Canyon rock daisy (*Perityle congesta***)**, a small, rounded perennial with a more delicate appearance than Emory's rock daisy, has tiny, yellow disk flowers and no ray flowers. It thrives from river to rim, clinging to rock crevices. Grand Canyon rock daisy can be confused with the **white-flowered arrowleaf (***Pleurocoronis pluriseta***)**, which also bears only disk flowers and small, arrowlike leaves.

Emory rock daisy honors Major William Hemsley Emory (1811–1887), a cartographer and naturalist who surveyed the states that shared boundaries with Mexico following the Mexican War. He fostered many botanists under his command.

Desert-Dandelion *Malacothrix glabrata*
—Photograph © 2006 by Kristin Huisinga
Inset —Photograph © 2006 by Kristin Huisinga

White-Flowered Arrowleaf *Pleurocoronis pluriseta*
—Photograph © 2006 by Lisa A. Hahn

Emory Rock Daisy *Perityle emoryi*
—Photograph © 2006 by Kristin Huisinga

Grand Canyon Rock Daisy *Perityle congesta*
—Photograph © 2006 by Kristin Huisinga

Pygmy-Cedar *Peucephyllum schottii*
SUNFLOWER FAMILY (Asteraceae/Compositae)

Plants: Rounded shrub, to 2 meters tall. **Leaves:** Alternate, linear, succulent, 1 to 2 centimeters long. **Flowers:** Solitary; disk flowers yellow; ray flowers lacking. **Fruit:** Achene. **Flowering Season:** April to May. **Elevation:** 1,200 to 2,700 feet.

Pygmy-cedar, an aromatic shrub, looks like a bonsai version of a needled fir tree. However, it is not a conifer at all, even though its simple, needlelike leaves and pinelike aroma might lead you to guess otherwise. To identify it as a member of the sunflower family, look for the hairy, silvery achenes before they blow away, leaving behind the dried, fully opened base of the flower heads.

Pygmy-cedar occurs in California, Nevada, Utah, and Arizona. In the Grand Canyon, it grows from Waltenberg Canyon (downstream of Bass Rapids), downstream to Grand Wash Cliffs. Frost damages its leaves so it often grows in the warmest microhabitats such as directly out of the rock layers at river level and in gravelly washes and dry slopes.

The genus name, *Peucephyllum*, is Greek for "fir-leaf," referring to its conifer-like leaves. The epithet, *schottii*, honors Austrian botanist Heinrich Wilhelm Schott (1794–1865), who was known for his work with species in the family Araceae, home to many houseplants.

Arrow-Weed *Pluchea sericea*
SUNFLOWER FAMILY (Asteraceae/Compositae) (*Tessaria sericea*)

Plants: Upright, many stemmed shrub, to 3 meters tall. **Leaves:** Alternate, lanceolate, silvery green, 1 to 5 centimeters long. **Flowers:** In terminal panicles; disk flowers pink to purple; ray flowers lacking. **Fruit:** Achene with bristly pappus. **Flowering Season:** March to June. **Elevation:** 1,200 to 3,200 feet.

Arrow-weed, a common riparian shrub, forms large colonies on beaches and cobblebars. Its silvery green leaves and long straight branches set it apart from other plants in the river corridor. Shortly after the completion of Glen Canyon Dam, arrow-weed took up new residence on open sandy areas that no longer received scouring floods. Over time, the arrow-weed patches became choked with the dead wood and grew slowly. An experimental flood in 1996 buried arrow-weed plants in sediment more than 1 meter deep, and the swift flows removed old stems. Vigorous new stems emerged several weeks later from the buried roots, forming healthier thickets. The experiment demonstrated that regular floods benefit arrow-weed communities.

Arrow-weed occurs in riparian areas throughout the Southwest and into Mexico. In the Grand Canyon, it grows the length of the river.

Native American people use the stiff, straight branches for arrow shafts, basketry, and thatch roofing material. The flowers produce nectar that is a source of honey. The specific epithet, *sericea*, describes the straight silky hairs covering the leaves and stems.

Pygmy-Cedar *Peucephyllum schottii* —Photograph © 2006 by Roger Dale

Pygmy-Cedar *Peucephyllum schottii*
—Photograph © 2006 by Mary Allen

Arrow-Weed *Pluchea sericea*
—Photograph © 2006 by Lisa A. Hahn

Poreleaf, Poreweed
Porophyllum gracile
SUNFLOWER FAMILY (Asteraceae/Compositae)

Plants: Erect perennial herb, to 70 centimeters tall. **Leaves:** Alternate or opposite, filiform to linear, bluish green, 1 to 5 centimeters long. **Flowers:** Solitary or in cymes; disk flowers white to lavender; ray flowers lacking. **Fruits:** Achene with bristly pappus. **Flowering Season:** March to November. **Elevation:** 1,200 to 5,400 feet.

The pungent odor permeating the air around poreleaf may be your first introduction to this slender-stemmed, bushy plant. While pleasing to some, others find the smell offensive. Visible to the naked eye, the sunken glands on the leaves and flower heads emit this characteristic fragrance, which protects the plant by repelling insects and deterring grazing animals. However, deer feed on poreleaf, likely leading to another common name, *yerba del venado*, which means "deer herb."

Poreleaf occurs on rocky slopes and dry washes in desertscrub communities from Baja California to Texas. In the Grand Canyon, this plant is a staple of the Inner Canyon flora, occurring the length of the river.

Indigenous people in the southwestern United States and northern Mexico used poreleaf for a wide variety of medicines to treat aches and pains, malaria, diarrhea, and urinary tract disorders. The genus name, *Porophyllum*, is from the Greek *poros*, meaning "passage," and *phyllon*, meaning "leaf," descriptive of the glands that exude oil.

Cudweed
Pseudognaphalium stramineum
SUNFLOWER FAMILY (Asteraceae/Compositae)
(*Gnaphalium chilense*)

Plants: Annual or biennial herb, 15 to 80 centimeters tall. **Leaves:** Alternate, linear to oblong, reduced along stem, 1.5 to 7 centimeters long. **Flowers:** In panicles, with yellow bracts; disk flowers white to yellow; ray flowers lacking. **Fruit:** Achene with bristles. **Flowering Season:** May to October. **Elevation:** 2,000 to 3,100 feet.

Cudweed is clothed in dense, woolly hairs that make the leaves and stems appear soft and gray. Like many members of the sunflower family, the tightly packed flower heads are crowded at the stem tips.

Cudweed grows from British Columbia throughout the western United States to Mexico and in scattered states east of the Mississippi River. In the Grand Canyon, it grows along the river, at seeps and springs, and in side canyons the length of the river.

Another equally common cudweed species in the Inner Canyon is **Wright cudweed** (*Pseudognaphalium canescens* **ssp.** *canescens*; synonym **Gnaphalium wrightii**), which is usually a taller, more highly branched perennial with white flower bracts. It is easy to confuse the cudweeds with another woolly member of the sunflower family, the ground-hugging **desert nest-straw (*Stylocline micropoides*)**. Desert nest-straw produces copious amounts of minuscule, cottonball-like fruits that are often clustered around ant hills where ants have gathered them. This species occupies a completely different niche on dry, open slopes and flats.

The old genus name, *Gnaphalium*, is from the Greek *Gnaphalion*, meaning "lock of wool," referring to the dense white hairs covering cudweed.

Poreleaf *Porophyllum gracile* —Photograph © 2006 by Kate Watters

Cudweed *Pseudognaphalium stramineum*
—Photograph © 2006 by Glenn Rink

Developing flower heads of Desert Nest-Straw
Stylocline micropoides
—Photograph © 2006 by Max Licher

Mature fruits of Desert Nest-Straw
Stylocline micropoides
—Photograph © 2006 by Kate Watters

Desert Chicory
Rafinesquia neomexicana

SUNFLOWER FAMILY (Asteraceae/Compositae)

Plants: Winter annual herb, to 40 centimeters tall. **Leaves:** Alternate, elliptic to oblanceolate, toothed to narrowly lobed, 3 to 15 centimeters long. **Flowers:** Solitary or in few-flowered panicles; disk flowers lacking; ray flowers white. **Fruit:** Achene with bristly pappus. **Flowering Season:** March to May. **Elevation:** 1,200 to 4,200 feet.

In early spring, the showy, bright white flower heads of desert chicory peak out from the skeletons of dormant desert shrubs. The woody shrubs lend support to desert chicory's weak stems and may also provide shelter from the elements and herbivores. Each ray flower is fringed, as if cut with pinking shears, and its stems leak milky sap when broken.

Desert chicory occurs in California, Nevada, and Utah, east to Texas, and into Mexico. In the Grand Canyon, it grows on desert slopes, often nursed beneath larger plants, from the Tanner area to Grand Wash Cliffs.

Desert chicory resembles the European chicory (*Cichorium intybus*), which is commonly used as a coffee substitute because of its rich flavor. The genus name, *Rafinesquia*, honors Constantine Samuel Rafinesque-Schmaltz (1783–1840), a nineteenth-century botanist and friend of John James Audubon.

Sow-Thistle
Sonchus species

SUNFLOWER FAMILY (Asteraceae/Compositae)

Plants: Exotic annual herb, to 1 meter tall. **Leaves:** Alternate, entire or lobed, with toothed margins, clasping at the base, 3 to 30 centimeters long. **Flowers:** In panicles; disk flowers lacking; ray flowers yellow. **Fruit:** Achene with bristly pappus. **Flowering Season:** February to October. **Elevation:** 1,200 to 3,200 feet.

Sow-thistle is a prominent weed in moist soils, along rivers, side creeks, and in seeps and springs. Its stems and prickly foliage exude a milky sap, and its yellow flower heads bear a striking resemblance to common dandelion (*Taraxacum officinale*). To eradicate this non-native species from Grand Canyon National Park would be an almost impossible task, given its widespread distribution and the ability of each flower head to produce up to 250 seeds.

The two primary species of sow-thistle in the Grand Canyon are very difficult to distinguish without fruit and a hand lens. **Common sow-thistle (*Sonchus oleraceus*),** a daintier plant, typically has deeply lobed leaves, while the robust **spiny sow-thistle (*S. asper*)** has entire leaves with prickly margins. However, leaf shape varies widely and these species may hybridize. The key difference is in the ribbed fruit. Common sow-thistle has thinner, wrinkled achenes and spiny sow-thistle has somewhat wider, flattened achenes. Look for the fewer-flowered, and thus narrower, heads to distinguish **wild lettuce (*Lactuca serriola*)** from sow-thistle. They all have milky sap.

Sow-thistles are native to Europe, but are fairly common in disturbed habitats throughout the world. In the Grand Canyon, both sow-thistles are found the length of the river.

Sow-thistles boast a multitude of medicinal properties. The genus name, *Sonchus*, is Greek for "sow-thistle," alluding to its ancient status in the world's flora.

Desert Chicory *Rafinesquia neomexicana* —Photograph © 2006 by Dan Hall

Common Sow-Thistle *Sonchus oleraceus*
—Photograph © 2006 by Kristin Huisinga

Spiny Sow-Thistle *Sonchus asper*
—Photograph © 2006 by Glenn Rink

Wire Lettuce *Stephanomeria pauciflora*
SUNFLOWER FAMILY (Asteraceae/Compositae)

Plants: Many branched, perennial herb, 30 to 60 centimeters tall. **Leaves:** Alternate, variously lobed, scalelike, 3 to 7 centimeters long. **Flowers:** Solitary or in panicles; disk flowers lacking; ray flowers pink to lavender. **Fruit:** Achene with feathery bristles. **Flowering Season:** February to November. **Elevation:** 1,200 to 7,900 feet.

Wire lettuce, a thin-stemmed, shrubby plant with delicate pinkish purple flowers, lives on dry slopes along the Colorado River. On first glance, this herb may remind you of other common, sparsely leaved shrubby plants with tangled branches, such as chuckwalla's delight (*Bebbia juncea* var. *aspera*) and spiny aster (*Chloracantha spinosa*). When in doubt, break a small piece of the stem to reveal the characteristic milky sap of wire lettuce.

Wire lettuce occurs from California east to Kansas, Oklahoma, and Texas, and south to Mexico. In the Grand Canyon, it grows from river to rim on cobble bars, desert flats, and dry slopes.

Two other species of wire lettuce are abundant in the Inner Canyon. They are extremely easy to confuse given their leafless appearance and similar flowers. Another perennial species of wire lettuce (***Stephanomeria tenuifolia***) is more slender and less branched than *S. pauciflora*, growing more often at higher elevations. The only easy way to tell the two perennial species apart is to examine their fruits. The achene of *S. pauciflora* has brown-tinged bristles that are plumose only from the middle to the tip and *S. tenuifolia* has bright white, plumose bristles from tip to base. **White-plume wire lettuce (*S. exigua*)**, an annual with a small taproot, appears more fragile than both perennial species despite its dense branching.

Some believe that because of its milky sap, wire lettuce increases milk flow in lactating mothers when taken internally. The root was used as chewing gum, as an ingredient for paint, and as a narcotic.

Fetid-Marigold, Dogweed *Thymophylla pentachaeta* var. *pentachaeta*
SUNFLOWER FAMILY (Asteraceae/Compositae) (*Dyssodia pentachaeta* var. *pentachaeta*)

Plants: Perennial herb, 10 to 30 centimeters tall. **Leaves:** Opposite, needlelike, pinnately divided, to 2 centimeters long. **Flowers:** Solitary or in cymes, stalked; disk and ray flowers yellow. **Fruit:** Achene with scaly pappus. **Flowering Season:** March to December. **Elevation:** 1,200 to 5,000 feet.

The common name *fetid-marigold* is a misnomer for this cheerful little perennial. Its pleasant, aromatic fragrance, which distinguishes it from many other yellow-flowered members of the sunflower family, originates from conspicuous oil glands embedded in the leaves, stems, and the margins of the flower bracts. You may find this attractive plant blooming almost year-round, even when few other plants are in flower.

Fetid-marigold occurs from California east to Texas and south to northern Mexico. A disjunct population occurs as far away as Argentina. In the Grand Canyon, it grows the length of the river extending as high as the Tonto Plateau and the Esplanade.

Commonly confused with snakeweed (*Gutierrezia* species), fetid-marigold has many more ray flowers and larger heads. Another species, **needleleaf dogweed (*Thymophylla acerosa*)**, has linear leaves and unstalked flower heads, and grows mostly in the western part of the Grand Canyon above 3,200 feet. A rayless relative, **San Felipe dyssodia (*Adenophyllum porophylloides*)**, is more limited in distribution and prefers schist and limestone substrates.

Wire Lettuce *Stephanomeria pauciflora*
—Photograph © 2006 by Harlan Taney

Wire Lettuce *Stephanomeria tenuifolia*
—Photograph © 2006 by Max Licher

Fetid-Marigold *Thymophylla pentachaeta* var. *pentachaeta* —Photograph by Lori Makarick/National Park Service

Trixis
Trixis californica

SUNFLOWER FAMILY (Asteraceae/Compositae)

Plants: Woody evergreen shrub, to 2 meters tall. **Leaves:** Alternate, lanceolate, dark green, entire or toothed, 2 to 12 centimeters long. **Flowers:** In cymes; disk flowers yellow; ray flowers lacking. **Fruit:** Achene with prominent bristles. **Flowering Season:** February to August. **Elevation:** 1,200 to 6,400 feet.

The beautiful, whitish stems of trixis reflect the sun's rays, thereby reducing heat and reserving moisture, a common desert plant adaptation. The gland-dotted leaves emit a spicy odor when crushed. Although its numerous seeds are well adapted for wind dispersal and float through the air like dandelion seeds, germination appears to be infrequent.

When trixis lacks flowers, its identification can stump the most practiced botanist. Its shiny, sharp-toothed leaves resemble those of the wide-leaved *Baccharis* species, but the more compact trixis typically grows in drier locations. Long after the plant has flowered, look for the old flower stalks with chalky white bracts that linger into the fall.

Trixis occurs from California, east to Texas, and south to Baja California and Sonora, Mexico. In the Grand Canyon, it prefers open, rocky slopes from River Mile 76 to Grand Wash Cliffs, with scattered distributions farther upriver. The Grand Canyon plants represent the northern limit of this typically Sonoran Desert plant. The Colorado River probably serves as a corridor for the eastward dispersal of trixis, which is also apparent in other species, including California barrel cactus (*Ferocactus cylindraceus*) and beargrass (*Nolina microcarpa*).

Native people use trixis for tobacco and medicine.

Cocklebur
Xanthium strumarium

SUNFLOWER FAMILY (Asteraceae/Compositae)

Plants: Annual herb, 15 to 150 centimeters tall. **Leaves:** Alternate, round or cordate, yellowish green, often lobed, 2 to 12 centimeters long. **Flowers:** In axillary clusters, male and female separate; disk and ray flowers inconspicuous. **Fruit:** Spiny bur. **Flowering Season:** April to October. **Elevation:** 1,700 to 4,000 feet.

Cocklebur is easy to identify because of its lush, green leaves and prominent, golden brown burs. Plants benefit from river fluctuations and floods, which create newly disturbed soil and disperse seeds. Burs can survive for many days floating in water. Once established in a new area, a single plant can multiply quickly. The male flowers reside above the females, and a gust of wind causes pollen to rain down onto the females. One healthy plant can produce more than five hundred seeds, which not only drop near the plant but also cling to clothing or fur of unsuspecting passersby.

Sacred datura (*Datura wrightii*) is another stout plant with a large, burlike fruit but its tubular, white flowers and dark green leaves distinguish it from cocklebur. Without flowers or fruit, young sunflower plants (*Helianthus annuus*) resemble cocklebur.

Cocklebur occurs worldwide although its country of origin is unknown. In the United States, it grows in every state except Alaska. In the Grand Canyon, it prefers open, disturbed areas the length of the river.

Cocklebur's poisonous properties contribute to its status as a noxious weed in many areas, but it also has medicinal properties. Many people who are allergic to ragweed suffer similar effects when exposed to cocklebur pollen.

Trixis *Trixis californica*
—Photograph © 2006 by Arizona Raft Adventures

Trixis *Trixis californica*
—Photograph © 2006 by Kathy Darrow

Cocklebur *Xanthium strumarium* —Photograph © 2006 by Kate Watters

Mojave Aster
Xylorhiza tortifolia
SUNFLOWER FAMILY (Asteraceae/Compositae)
(*Machaeranthera tortifolia*)

Plants: Perennial herb or small shrub, to 75 centimeters tall. **Leaves:** Alternate, linear to elliptic, spiny-margined, 1 to 10 centimeters long. **Flowers:** Solitary on long stalks; disk flowers yellow; ray flowers white to lavender. **Fruit:** Achene with bristly pappus. **Flowering Season:** February to June. **Elevation:** 1,200 to 6,100 feet.

In the spring, copious pinkish white flowers cover the mounded foliage of Mojave aster. Large plants may produce fifty flowers at one time. When Mojave aster lacks flowers, it resembles spiny brickellbush (*Brickellia atractyloides*) and the widespread goldenbush (*Isocoma acradenia*). Spiny brickellbush has triangular, leathery leaves with spine-tipped margins and prominent veins. Goldenbush has linear leaves with rounded teeth and it grows in drainage bottoms. Both lack the hairy, glandular surface of the larger-leaved Mojave aster.

Mojave aster occurs in the deserts of California, Utah, and Arizona. In the Grand Canyon, it grows on talus slopes and desert flats and in dry side canyons and other sunny places. In the westernmost parts of the canyon, it is often associated with creosotebush (*Larrea tridentata* var. *tridentata*).

Gray-dotted, black caterpillars armored with rows of arching spines feed on the leaves of Mojave aster. These are the larvae of a small butterfly, the desert checkerspot (*Chlosyne neumoegeni*). Adults drink the nectar and lay eggs on the plants. A small moth (*Schinia* species), also associated with Mojave aster, has lavender-brown wings that are easily camouflaged by the blooming flowers. The larvae of these moths feed on the developing seeds.

Lemon Verbena, Wright's Lippia
Aloysia wrightii
VERVAIN FAMILY (Verbenaceae)
(*Lippia wrightii*)

Plants: Shrub, to 2 meters tall. **Leaves:** Opposite, ovate, toothed, 4 to 17 millimeters long. **Flowers:** In terminal spikes, white, 1 to 4 millimeters long. **Fruit:** Nutlets. **Flowering Season:** August to October. **Elevation:** 1,800 to 4,500 feet.

Lemon verbena's miniature white flowers, arranged in slender spikes, emit a sweet fragrance that invites bees, butterflies, and visitors alike to take a closer look. Rub the small, thick leaves of lemon verbena between your fingers to release their musty, lemony odor, which is the best clue to this plant's identity. In the fall, lemon verbena blends into the landscape when it drops its velvety-bottomed leaves. It is a long-lived, drought-tolerant, and extremely hardy member of the Grand Canyon flora.

Lemon verbena occurs from California and Nevada east to western Texas and south into Mexico. While it is a characteristic plant of the Chihuahuan Desert, lemon verbena also inhabits the Mojave and Sonoran Deserts. It grows on open, rocky slopes in or above the creosotebush zone. In the Grand Canyon, it occurs from Silver Grotto downstream to Grand Wash Cliffs and at Toroweap Point.

Havasupai people dry the fragrant leaves to make a tasty beverage, and relatives of this species are used widely in Mexico as a spice. The genus name, *Aloysia*, celebrates Maria Luisa Teresa De Bourbon (1751–1819), wife of Charles IV of Spain. In 1853, Asa Gray, America's leading botanist in the mid-nineteenth century, named the species for Charles Wright (1811–1885), who first collected it in southern Arizona while working on a survey of the post–Mexican War boundary.

Mojave Aster *Xylorhiza tortifolia* —Photograph © 2006 by Lisa A. Hahn

Lemon Verbena *Aloysia wrightii*
—Photograph © 2006 by Celia Southwick

Lemon Verbena *Aloysia wrightii*
—Photograph © 2006 by Max Licher

Notch-Leaf Scorpionweed *Phacelia crenulata*
WATERLEAF FAMILY (Hydrophyllaceae)

Plants: Annual herb, to 80 centimeters tall. **Leaves:** Alternate, narrowly elliptic to oblong, variously lobed, to 10 centimeters long. **Flowers:** In scorpioid cymes, blue to violet, to 1 centimeter long. **Fruit:** Capsule with boat-shaped, bicolored seeds. **Flowering Season:** February to September. **Elevation:** 1,200 to 5,000 feet.

Notch-leaf scorpionweed, a robust annual, has tightly packed, sessile, purple blossoms on a strongly coiled flowering stalk. Prior to flowering, the flattened basal rosettes dot canyon bottoms. The leaves and stems have gland-tipped hairs that exude a pungent odor.

There are approximately 150 species of *Phacelia*, all native to the New World and best represented in the western United States and northern Mexico. In the Grand Canyon, twenty species are distributed from the river to both rims. Notch-leaf scorpionweed, one of the more common species in the Grand Canyon, includes at least five varieties. Identifying individual species of scorpionweeds requires a microscopic look at the seed shape.

Two other smaller, more delicate annual scorpionweeds are endemic to northern Arizona and have overlapping ranges in the Grand Canyon. **Pennyroyal-leaf scorpionweed (*P. glechomifolia*)** and **Grand Canyon scorpionweed (*P. filiformis*)** have rounded leaves bearing scalloped margins and stalked, pale pink to lavender blossoms that are widely spaced on gently curving flower stalks. To set them apart, look for the smaller flowers of Grand Canyon scorpionweed (5 to 8 millimeters long) compared to the larger flowers of pennyroyal-leaf scorpionweed (10 to 15 millimeters long).

Notch-leaf scorpionweed occurs from southern California to New Mexico, including Utah and Colorado. In the Grand Canyon, it grows on flats and slopes, and in washes the length of the river. The exceptionally robust, large-leaved annuals with uniformly colored seeds found along the Bright Angel and North Kaibab Trails are **caterpillar weed (*P. crenulata* var. *ambigua*)**.

The sticky, glandular hairs of *Phacelia* species contain phenolic compounds that, upon contact, can produce allergic responses, including skin rashes similar to those of poison-ivy. The genus name, *Phacelia*, is from the Greek word, *phakelos*, or "cluster," referring to the densely packed, coiled flowering stalks.

Notch-Leaf Scorpionweed *Phacelia crenulata* —Photograph © 2006 by Glenn Rink

Notch-Leaf Scorpionweed *Phacelia crenulata*
—Photograph © 2006 by Kathy Darrow

Grand Canyon Scorpionweed *Phacelia filiformis*
—Photograph © 2006 by Glenn Rink

Arizona Fiesta Flower *Pholistoma auritum* ssp. *arizonicum*
WATERLEAF FAMILY (Hydrophyllaceae)

Plants: Viny annual herb, to 60 centimeters long. **Leaves:** Lower opposite and upper alternate, oblong to lanceolate, deeply lobed and toothed, to 16 centimeters long. **Flowers:** Solitary or in cymes, blue to lavender. **Fruits:** Capsule with stout bristles. **Flowering Season:** February to April. **Elevation:** 1,200 to 3,600 feet.

At first glance, it might appear as if small, bluish lavender flowers are emerging from a dormant shrub. A closer look reveals that Arizona fiesta flower, a weak-stemmed, trailing vine, is climbing over a shrub, using it for support. Backward-facing, sharp bristles that clothe the vine's stems and leaves might grab your clothing.

Arizona fiesta flower occurs in California, Nevada, and Arizona. In the Grand Canyon, it grows on dry, rocky slopes and flats from Saddle Canyon to Grand Wash Cliffs.

Small-flowered eucrypta (*Eucrypta micrantha*), a related species in the waterleaf family, has extremely similar flowers but is smaller overall and lacks the viny habit of Arizona fiesta flower. It also lacks the bristles, emits a faint odor, and has sticky leaves with more rounded lobes.

The genus name, *Pholistoma*, is from the Greek words *pholis*, meaning "scale" and *stoma*, meaning "mouth," describing the scales in the flower's throat. The specific epithet, *auritum*, means "eared," referring to the distinct clasping base of the leaves.

Arizona Fiesta Flower *Pholistoma auritum* ssp. *arizonicum*
—Photograph © 2006 by David Edwards

Small-Flowered Eucrypta *Eucrypta micrantha* —Photograph © 2006 by Glenn Rink

PLANT ANATOMY ILLUSTRATIONS

LEAF SHAPES

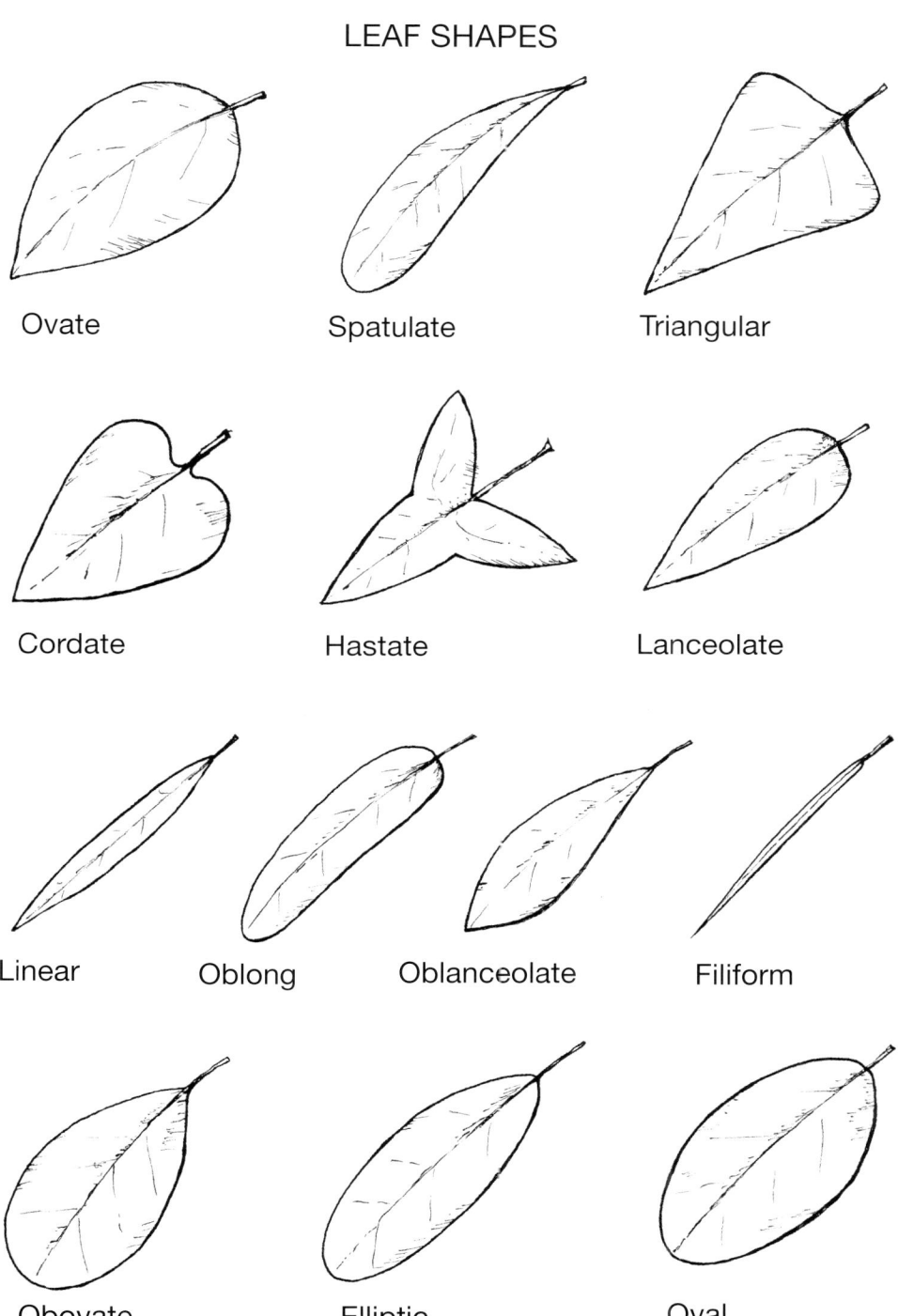

LEAF DIVISION—GENERAL

Simple

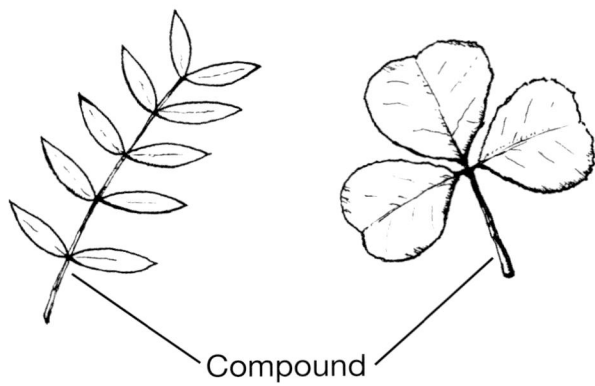
Compound

LEAF DIVISIONS

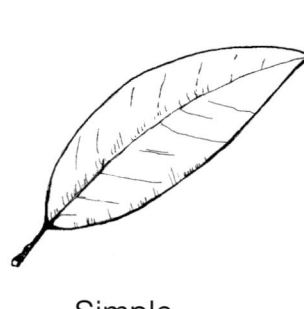
Simple

Palmate

Pinnate

Bipinnate or
Twice Pinnate

Tripinnate

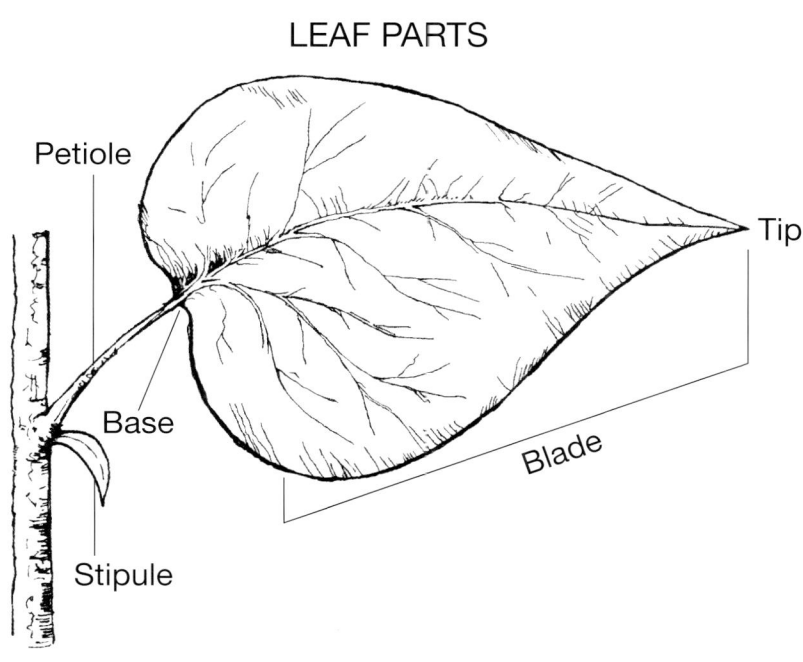

LEAF ARRANGEMENTS

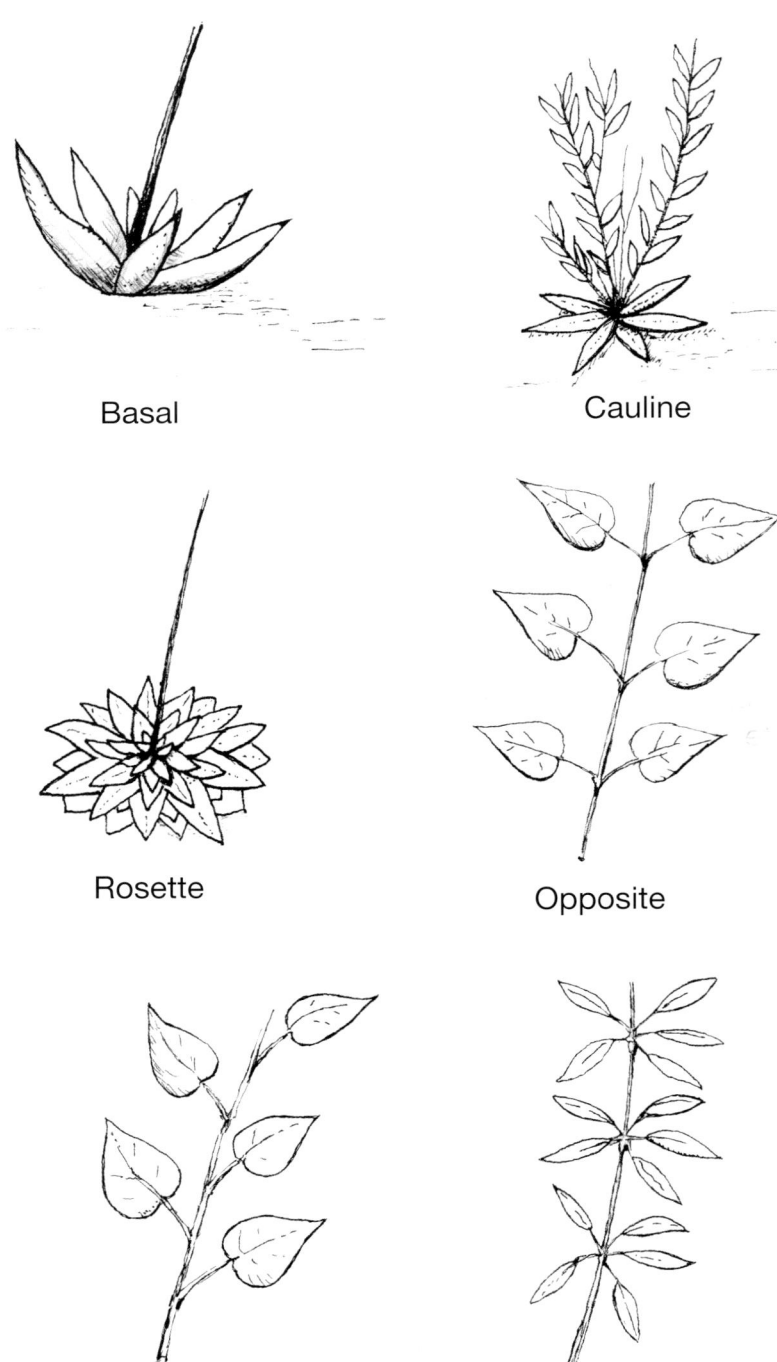

Basal

Cauline

Rosette

Opposite

Alternate

Whorled

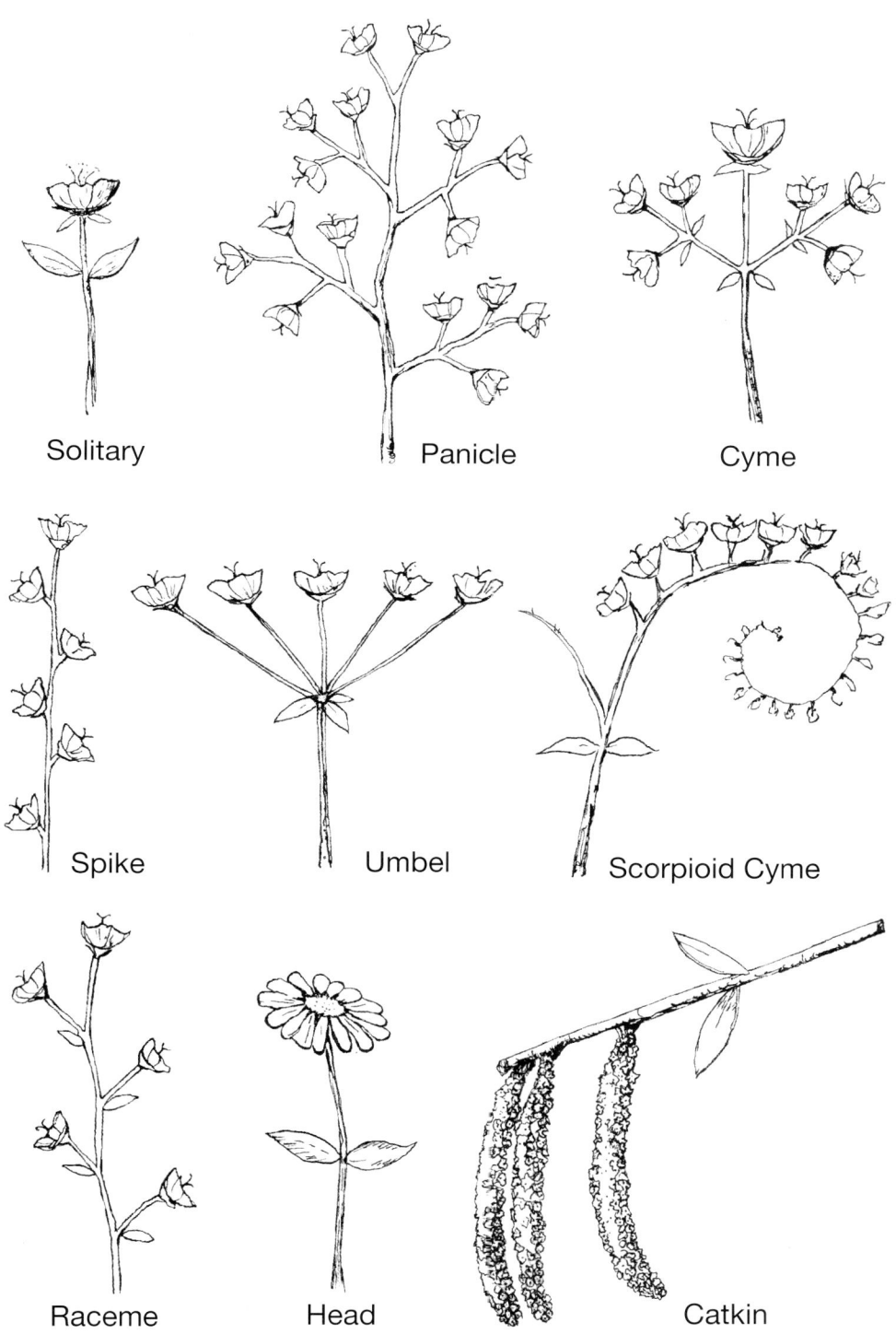

FLOWER PARTS

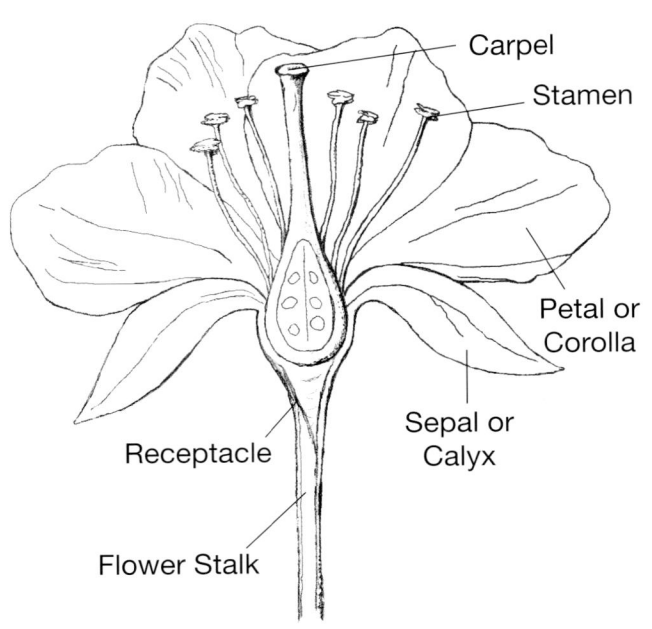

STAMEN

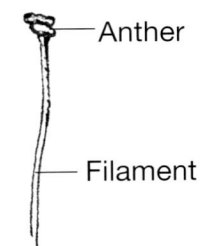

PISTIL OR CARPEL

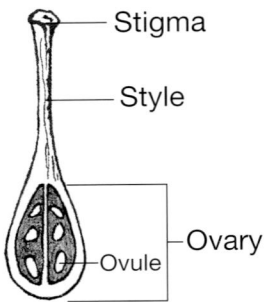

FLOWERS IN THE SUNFLOWER FAMILY

FLOWERS IN THE PEA FAMILY

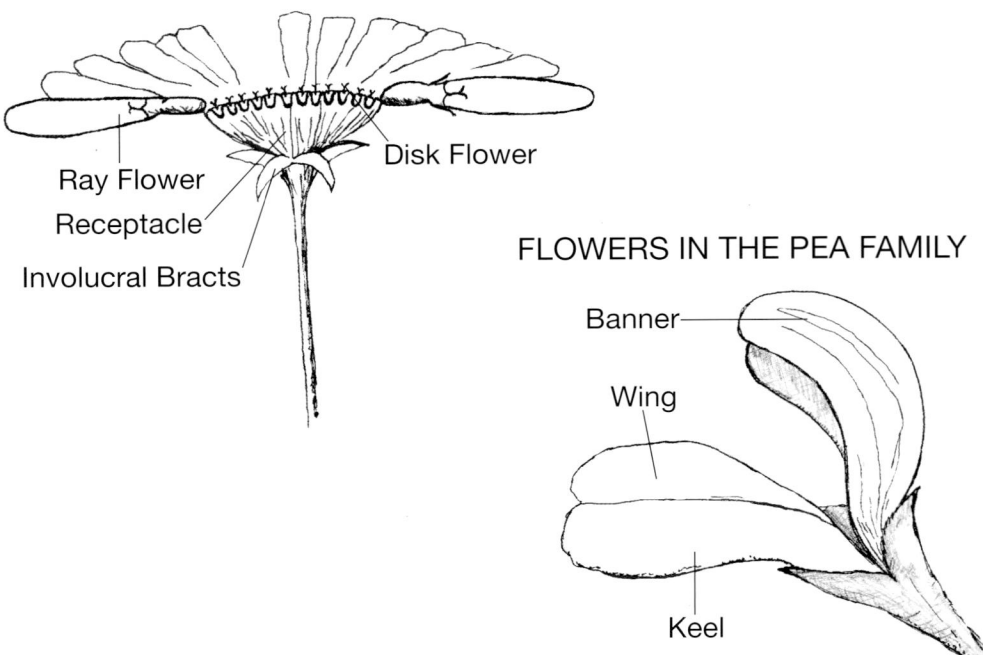

GLOSSARY

achene. A small, hard, one-seeded fruit that functions as a single seed; does not split open at maturity.

acorn. The fruit of an oak; hard, dry, and one-seeded with a cuplike base; does not split open at maturity.

aerial. Occurring above the surface of the ground.

algae. A diverse group of photosynthetic organisms that are largely aquatic and lack true roots, stems, and leaves; range in size from tiny single-celled forms to giant kelp.

alkaline. Said of soil or water with a high pH; not acidic.

alkaloid. A toxic nitrogen-containing substance, usually serving to defend the plant.

alternate. Leaf arrangement in which there is one leaf per node; not opposite.

annual. A plant that lives only one year; it grows from seed and produces seed in one year.

anther. The part of the stamen (male sex organ) that produces pollen.

areole. A structure in cacti that bears spines or glochids or both.

astringent. A substance that constricts body tissues; contracting.

awn. A bristlelike part, typically borne at the tip of some other plant part.

axil. The upper angle formed by a leaf with the stem.

axillary. Describes a flower that forms in the axil of a leaf and stem.

bacteria. A diverse group of microscopic, nonphotosynthetic organisms.

banner. Upper petal of a pea flower.

barb. A short, stiff, hairlike structure that is bent backward or downward.

bark. Outer layer of a woody stem.

basal. Leaves arranged at the base of the stem at or near ground level.

beak. A prolonged, typically narrowed tip of some structure, such as a fruit.

berry. A fleshy fruit, typically with several seeds.

biennial. A plant that lives two years, the first year producing leaves and a thick taproot, the second year developing an erect stem with flowers.

bipinnate. Pinnately divided two times.

blade. The flat part of a leaf or petal.

blossom. A flower.

bract. A small, modified, often pigmented leaf typically situated at the base of a flower or flower cluster.

bristle. A stiff, hairlike structure.

bulb. A thickened, fleshy structure that typically forms belowground and functions in food storage and reproduction.

bunchgrass. A perennial grass that grows in clumps; not matted.

bur. A fruit covered with hooked spines, which attach to animals and aid in seed dispersal.

calyx. A collective term for all the sepals.

capillary. Hairlike; very fine and slender.

capsule. A fruit that dries and splits open at maturity, shedding its seeds; typically contains two or more compartments.

catkin. A dangling, spiked inflorescence with densely arranged, unisexual flowers having no petals; commonly found in the willow family.

cauline. Along the stem.

chaparral. A plant community characterized by dense thickets of evergreen shrubs.

chlorophyll. The green pigment that allows plants to photosynthesize, making their own food.

clasp. Said of some leaf blades that partly or totally encircle the stem.

clones. A group of genetically identical plants that have arisen from a single parent through vegetative reproduction.

community. A group of plants living together in a given habitat.

compound. Said of a leaf or inflorescence that has two or more parts.

cone. A structure involved in gymnosperm reproduction; in conifers, consists of a cylindrical or spherical cluster of spore- or seed-bearing scales emerging from a single axis.

corm. A short, swollen underground stem with thin, delicate leaves; functions in food storage and reproduction.

corolla. A collective term for a flower's petals.

cross-pollination. The transporting of pollen from one plant to another.

crown. The branched top section of a tree; the persistent foundation of a perennial herb.

culm. The stem of a grass or sedge.

cyanobacteria, blue-green algae. A group of generally blue-green primitive algae; some cyanobacteria fixate nitrogen.

cyme. A flat- or round-topped panicle in which the terminal flowers bloom first.

deciduous. A plant whose leaves fall at the end of the growing season. Called **drought deciduous** if leaves fall during periods of drought.

desertscrub. A dry environment characterized by widely spaced shrubs and succulents, such as creosotebush and cacti.

diffuse. Widely spread.

disjunct. Inhabiting widely scattered geographic areas; discontinuous.

disk flower. One of the central flowers on the head of a member of the sunflower family; a tubular-shaped flower typically with five teeth or lobes and lacking a ray (flattened extension).

divided. Said of leaves that are separated to the base (or to the midvein).

dominant. One of the most important plants of a given community because of numbers or size; a plant that has a major effect on other plants of the community.

dormant. A period of metabolic rest.

drupe. A fleshy fruit, often with one seed surrounded by a hard casing, such as a peach, plum or cherry; does not split open at maturity.

ecosystem. A community of plants and animals plus the physical environment that influences them.

endemic. Restricted to a particular geographic area.

entire. Said of a leaf with a smoothly continuous margin that is neither toothed nor lobed.

erect. Being in an upright position.

ethnobotany. The study of how people of a particular culture or place have used plants (and their landscapes) for various aspects, such as food, clothing, and medicine.

evergreen. Retaining green leaves throughout the winter.

exotic. A plant species that has been moved, either intentionally or unintentionally, outside the area where it naturally evolved. Also called **non-native**.

filament. The threadlike stalk of an anther.

filiform. Resembling a thread.

fleshy. Thick and succulent; juicy.

foliage. The leaves of a plant.

follicle. A dry, podlike fruit with a single compartment, which splits open at maturity.

frond. The leaf of a fern.

fruit. The ripened ovary with its seed(s).

fungi. A diverse group of multicellular organisms that reproduce via spore.

furrowed. Deep, broad channels along a particular structure, such as a tree trunk.

germination. The process of a seed maturing into a seedling or a spore maturing into a prothallus.

gland. An organ that secretes a resinous, often sticky material.

glandular. Having glands, sometimes on the tips of hairs.

glaucous. Having a pale color, such as bluish green in leaves.

glochid. A small, usually barbed hair or bristle, typical of the cactus family.

glume. A bract or pair of bracts at the base of a sedge or grass spikelet.

gymnosperm. A seed plant that has "naked" ovules; not enclosed in an ovary. Includes conifers and ephedras.

habit. General appearance of a plant.

habitat. The home of a given plant, unique in having a particular set of environmental conditions.

hastate. Arrow-shaped with basal lobes pointed outward.

head. A dense cluster of flowers that lack stalks; the inflorescence of a member of the sunflower family.

herb. A plant lacking a hard, woody stem.

herbivore. An animal that primarily eats plant material. The eating of plants is called **herbivory**.

hip. A fleshy structure composed of a swollen cup-shaped receptacle and numerous achenes; common in roses.

hybridize. To effectively cross plant species from the same genus or plant species from different genera.

inconspicuous. Said of flowers with small green or greenish yellow petals that blend with adjacent parts.

indigenous. Originating and living naturally in a particular region; native.

inflorescence. A flower cluster.

infusion. A solution that is made by steeping a substance in water or oil to extract certain properties.

Inner Canyon. The area from the canyon rims to the canyon bottom, including the river corridor.

invasive. An introduced plant species that is established and likely to cause environmental or economic damage.

involute. Said of a leaf with margins rolled inward toward the upper plane.

keel. The two lower, united, keel-shaped petals of the pea family.

lanceolate. Lance-shaped; elongated with the widest point below the middle.

larvae (larval). The early, wormlike form of an insect prior to metamorphosis.

lateral. Situated at the side.

leaflet. One of the leaflike segments of a compound leaf.

lichen. A slow-growing organism composed of a fungus and algae; the fungus provides the support and protection, and the algae provide energy via photosynthesis.

ligule. A strap-shaped part; as related to grasses and some sedges, a membranous projection found between the blade and sheath of a leaf; as related to members of the sunflower family, the flattened portion of a ray flower.

liverworts. A group of small, primitive, terrestrial plants that reproduce via spore and lack true stems, roots, and leaves.

lobed. Said of a leaf that is cut or dissected but not all the way to the midvein; compare with **divided**.

marginal. Relating to a margin or edge.

mat-forming (matted). Low, dense, and spreading horizontally; resembling a mat or carpet.

membranous. Thin and translucent.

mosses. A group of primitive terrestrial plants that reproduce via spore; true mosses are small and lack true roots, stems, and leaves; club mosses are larger and have true roots, small leaves, and some vascular tissue.

mucilage. A slimy, sticky substance.

nectar. A sweet solution secreted by flowering plants that attracts pollinating insects and birds.

niche. The position of a species or population within an ecosystem and how that species or population affects and responds to changing environmental factors.

nodding. Said of a flower that hangs face down.

node. A joint on a stem; the point where the leaves are borne.

nodule. A small, swollen structure.

noxious. Said of a plant that is harmful or destructive.

nurse plant. A plant that facilitates the growth and survival of other plants, often in the form of protection.

nutlet. A hard-shelled, one-seeded fruit; a small nut.

oblanceolate. Inversely lanceolate; elongated with the widest part between the middle and apex.

obovate. Inversely ovate; broadest at the tip and tapering toward the base.

opposite. Leaf arrangement in which there are two leaves per node; not alternate.

ovary. The seed-containing part of the flower; matures into a fruit.

ovate. Having an egg-shaped outline; broadest at the base and tapering toward the tip.

ovule. An immature seed.

palatable. Appetizing; good-tasting.

palmate (palmately compound). Shaped like the palm of the hand with extended fingers

panicle. A branched raceme.

pappus. Bristlelike or scalelike parts borne on the ovary or fruit of members of the sunflower family. The pappus often functions in seed dispersal.

parasite. A plant growing on and deriving nourishment from another living plant.

perennial. A plant that lives more than two years; it may die down to the roots each winter but sprouts the next spring.

petal. One of the segments of the inner whorl of flower parts, usually colored or showy.

petiole. Leaf stalk.

phenolic compounds. A group of compounds produced by many plants for numerous reasons, such as protection from disease; responsible for the bright colors of many fruits and vegetables.

photosynthesize. To convert carbon dioxide and water into carbohydrates using sunlight as an energy source.

pinnate (pinnately compound). Perpendicular projections from a central axis; typically applied to a compound leaf.

pistil. The central (female) part of the flower, containing the ovary, style, and stigma. Also called **carpel.**

plumose. Resembling a plume or feather.

pod. A dry fruit, especially of the pea family, that splits open at maturity.

pollen. The tiny grains that contain male reproductive cells of seed plants.

pollinate. To transport pollen from the anther to the stigma.

poultice. A soft mass, often heated and spread on a cloth, which is then applied to an inflamed body part to provide moist warmth, ease pain, or prevent infection.

prickle. A small, pointed outgrowth emerging from external stem tissues.

proboscis. An elongated, tubular feeding and sucking structure of various invertebrates, such as insects.

prostrate. Growing horizontally along the ground.

prune. To remove parts of a plant to improve growth and shape.

raceme. An elongate, unbranched flower cluster, each flower having a stalk.

radial (radiate). Relating to a flower, star-shaped; spreading from a central point.

ray flower. One of the outer flowers of a member of the sunflower family, which has a flattened, elongate, colorful extension.

receptacle. The tip of a flowering stalk on which the parts of a flower are borne.

reduced. Diminished or smaller.

relict species. A species that has survived in an environment that has undergone significant change.

resin. A sticky, sometimes aromatic substance secreted by plants for protection.

rhizome (rhizomatous). An underground stem that produces roots and upright branches (stems); an organ by which plants (such as camelthorn and horsetail) spread.

riparian. Living along the banks of rivers, streams, or other water sources.

root crown. The juncture between the root and stem; the crown of the root.

rosette. A cluster or whorl of leaves arising at or near ground level.

saline. Salty; having sodium salts, potassium salts, or other alkali salts.

samara. A dry, winged fruit that does not split open at maturity.

scale. A thin, papery bract.

scorpioid. As related to a cyme, shaped like a scorpion's tail.

seed. A fertilized and ripened ovule.

semiaquatic. A plant that is adapted to live partly in water and partly on land.

sepal. One of the bractlike segments of the outer whorl of flower parts, usually green.

sessile. Not having a stalk; for example, a leaf without a petiole.

sheath. A tubular structure that encases, at least partly, another structure; for example, the base of a grass leaf that clasps the stem.

shrub. A woody plant that branches at or near ground level.

silicle. A dry fruit, with a length typically less than three times its width, that splits open at maturity.

silique. A dry fruit, characteristic of the mustard family, usually three times as long as wide; splits open at maturity.

simple. Said of a single, undivided leaf. Not compound.

sori. A group of spore-producing sacs found on the underside of a fern leaf.

spike. An elongate flower cluster with nonstalked flowers.

spikelet. A flower cluster found on grasses and sedges consisting of one or more flowers enclosed from below by one or a pair of glumes.

spine. A slender, stiff, sharp outgrowth emerging from below the epidermis; often a modified leaf or stipule. In cacti, the **central spine** is situated at the center of the areole. A **radial spine** radiates outward from the areole.

spore. A reproductive cell produced by ferns, fern allies, and fungi.

spur. A hollow extension of a petal or sepal, often containing nectar.

stamen. The part of a flower that contains pollen.

stigma. The part of a pistil that receives pollen.

stipule. Leaflike or bractlike part attached at the base of the petioles of some leaves. Stipules normally come in pairs and may be modified into spines.

stomate. A pore in leaves and green stems through which gases pass.

stolon. A horizontal stem that extends along the ground and roots at the nodes or tip, from which new plants grow.

strobilus (pl. strobili). A conelike structure of spore-bearing scales; a cone.

style. The narrow portion of the pistil, connecting the ovary with the stigma.

substrate. A surface on which an organism attaches or grows.

subterminal. Located near the tip.

succulent. Soft and juicy; filled with water.

talus. A sloping deposit of large, sharp angular rocks, often found at the base of a steep slope or cliff.

taproot. An elongate, unbranched, vertical root, like a carrot.

tendril. A threadlike modified leaf or stem used by twining plants for support.

terminal. Located at the tip.

thorn. A modified stem with a pointed tip.

translucent. Semitransparent; letting light pass through diffusely.

tree. A woody, usually tall plant, often having a trunk and branched crown.

tripinnate. Pinnately divided three times.

tuber (tuberous). A swollen underground stem or root that functions in food storage.

tubercle. A small prominence, such as on cacti stems.

tubular. Having a cylindrical and hollow form.

tufted. Densely clustered.

umbel. An umbrella-shaped flower cluster or inflorescence.

utricle. A small, one-seeded bladderlike fruit.

vascular. Pertaining to the conductive tissue that distributes water and nutrients within a plant.

vegetative reproduction. Asexual reproduction; spreading via rhizome or stolon rather than by seed.

vein. A cluster of vascular tissue that is often visible in leaves.

viable. Said of seeds that are able to germinate.

whorl. A group of three or more leaves, flowers, or petals radiating from a single point, such as from a node.

wing. A thin extension from a seed, fruit, or stem. One of a pair of lateral petals in a flower of the pea family.

woodland. An area dominated by widely spaced trees of low stature; savanna-like.

xeriscape. A landscaping method that utilizes drought-resistant plants in an effort to conserve water.

BIBLIOGRAPHY

Anderson, E. 2001. *The Cactus Family*. Portland, Ore.: Timber Press.

Anderson, J. 1999. *Plant Rarity in Arizona*. Tucson: The Plant Press. Arizona Native Plant Society.

Anderson, M. K. 1997. From Tillage to Table: The Indigenous Cultivation of Geophytes for Food in California. *Journal of Ethnobiology* 17(2):149–69.

Arizona Game and Fish Department. 2005. (*Flaveria mcdougallii*). Unpublished abstract compiled and edited by the Heritage Data Management System, Arizona Game and Fish Department, Phoenix.

Arizona Rare Plant Committee. 2002. *Arizona Rare Plant Field Guide*. A collaboration of agencies and organizations.

Ayers T. J., R. W. Scott, L. E. Stevens, K. Warren, A. M. Phillips III, and M. D. Yard. 1994. Additions to the flora of Grand Canyon National Park—I. *Journal of the Arizona-Nevada Academy of Science* 28:70-75.

Baldwin, B. G., S. Boyd, B. J. Ertter, R. W. Patterson, T. J. Rosati, and D. H. Wilken, eds. 2002. *The Jepson Desert Manual: Vascular Plants of Southeastern California*. Berkeley: University of California Press.

Benson, L. 1969. *The Cacti of Arizona*. 3rd ed. Tucson: University of Arizona Press.

Betancourt, J., T. Van Devender, and P. Martin. 1990. *Packrat Middens. The Last 40,000 Years of Biotic Change*. Tucson: University of Arizona Press.

Boatman's Quarterly Review. Volume 10 (4). 1997. "Then and Now," interview with Lois Jotter Cutter.

Bowers, J. E. 1999. *Flowers and Shrubs of the Mojave Desert*. Tucson, Ariz.: Southwest Parks and Monuments Association.

———. 1993. *Shrubs and Trees of the Southwest Deserts*. Tucson, Ariz.: Western National Parks Association.

Brasher, J. 1998. Celastraceae. *Journal of the Arizona-Nevada Academy of Science* 30:57-59.

Brian, N., W. Hodgson, and A. Phillips III. 1999. Additions to the flora of the Grand Canyon region—II. *Journal of the Arizona-Nevada Academy of Science* 32(2):117-28.

Brian, Nancy. 2000. *A Field Guide to the Special Status Plants of Grand Canyon National Park*. Grand Canyon, Ariz.: Science Center, Grand Canyon National Park.

California Plant Names: Latin and Greek Meanings and Derivations. A Dictionary of Botanical Etymology. http://www.calflora.net/botanicalnames/index.html.

Carothers, S. W., and S. W. Aitchinson. 1976. *An Ecological Survey of the Riparian Zones of the Colorado River between Lees Ferry and Grand Cliffs: Final Report, Grand Canyon National Park*. Grand Canyon, Ariz.: Colorado Research Series, Contribution Number 38.

Carothers, S. W., and B. T. Brown. 1991. *The Colorado River through Grand Canyon: Natural History and Human Change*. Tucson, University of Arizona Press.

Carpenter, A. T., and T. Murray. 1998. Element Stewardship Abstract: *Tamarix ramosissima, Tamarix pentandra, Tamarix chinensis, Tamarix parviflora, Tamarix gallica*. San Francisco, Calif.: The Nature Conservancy.

Castetter, E. F. 1935. Ethnobiological Studies in the American Southwest I: Uncultivated Native Plants Used as Sources of Food. *University of New Mexico Bulletin* 4(1):1-44.

Center for Sustainable Environments (CSE). 2002. *Safeguarding the uniqueness of the Colorado Plateau: An ecoregional assessment of biocultural diversity*. Flagstaff: Northern Arizona University.

Clark, C. 2000. *Encelia* and Its Relatives. http://www.csupomona.edu/~jcclark/encelia/species.html.

Clark, C. 1998. Phylogeny and adaptation in the *Encelia* alliance (Asteraceae: Heliantheae). *Aliso* 17(2):87-98.

Clary, K. 1997. *Phylogeny, Character Evolution, and Biogeography of Yucca L. (Agavaceae)*. Ph.D. Dissertation. University of Texas at Austin.

Clover, E. U., and L. Jotter. 1941. Cacti of the Canyon of the Colorado River and Tributaries. *Bulletin of the Torrey Botanical Club* 68:409-28.

———. 1944. Floristic Studies in the Canyon of the Colorado and Tributaries. *The American Midland Naturalist* 32:591-642.

Cornet, B. 1996. A new gnetophyte from the Late Carnian (Late Triassic) of Texas and its bearing on the origin of the angiosperm carpel and stamen. In *Flowering Plant Origin, Evolution and Phylogeny*, edited by D. W. Taylor and L. J. Hickey, 32-67. New York: Chapman & Hall.

Cronquist, A., A. H. Holmgren, N. H. Holmgren, J. L. Reveal, P. K. Holmgren, and R. C. Barneby. 1977–97. *Intermountain Flora, Vascular Plants of the Intermountain West*. Vols. 1, 3–6. Bronx, N.Y.: The New York Botanical Garden.

Desert USA. 2003. Four-wing Saltbush. http://www.desertusa.com/mag00/sep/papr/sitbush.html (accessed February 12, 2003).

Diver, Steve. ATTRA - National Sustainable Agriculture Information Service. http://www.attra.ncat.org/attra-pub/chinesewaterchestnut.html.

Duke, J. Holy Camelthorn (*Alhagia maurorum*) and Unholy Smallpox. http:/ www.21stcenturyradio.com/articles/1018041.html.

Dunmire, W. W., and G. D. Tierney. 1997. *Wild Plants and Native People of the Four Corners*. Santa Fe: Museum of New Mexico Press.

Elmore, F. H. 1976. *Shrubs and Trees of the Southwest Uplands*. Tucson, Ariz.: Southwest Parks and Monuments Association.

Epple, A. O., and L. E. Epple. 1995. *A Field Guide to the Plants of Arizona*. Helena, Mont.: Falcon Publishing, Inc.

Felger, R., and M. Moser. 1985. *People of the Desert and Sea: Ethnobotany of the Seri Indians*. Tucson: University of Arizona Press.

Flora of North America Editorial Committee. 1993. *Flora of North America: North of Mexico.* Volume 2: *Pteridophytes and Gymnosperms.* New York: Oxford University Press.

Flora of North America Editorial Committee. 2002. *Flora of North America.* Volume 26: *Liliidae.* New York: Oxford University Press.

Fowler, C. S. 1990. *Tule Technology: Northern Paiute Uses of Marsh Resources in Western Nevada.* Washington, D.C.: Smithsonian Institution Press

Gifford, E. W. 1936. Northeastern and Western Yavapai. *University of California Publications in American Archaeology and Ethnology* 34:247-345.

Gledhill, D. 2001. *The Names of Plants.* 3rd ed. Cambridge: Cambridge University Press.

Grand Canyon Natural History Association. 1936. *Check-List of Plants of Grand Canyon National Park.* Grand Canyon, Ariz.: Grand Canyon Natural History Association.

Grieve, M. 1971. *A Modern Herbal* (Volume I, A-H) New York: Dover Publications.

Hansen, M. 2002. *Sustainable Harvesting of* Petrophytum caespitosum *from Walnut Canyon National Monument.* Unpublished M.S. Thesis, Northern Arizona University.

Harris, J. G., and M. W. Harris. 2001. *Plant Identification Terminology: An Illustrated Glossary.* 2nd ed. Spring Lake, Utah: Spring Lake Publishing.

Hickey, M., and C. King. 2000. *The Cambridge Illustrated Glossary of Botanical Terms.* Cambridge: Cambridge University Press.

Hitchcock, A. S., and A. Chase. 1971. *Manual of the Grasses of the United States.* New York: Dover Publications, Inc.

Hodgson, W. C. 2001. *Food Plants of the Sonoran Desert.* Tucson: University of Arizona Press.

Hodgson, W. C. 2001. Taxonomic novelties in American *Agave* (Agavaceae). *Novon* 11(4):410-16.

Hogan, P., K. Huisinga, and K. Kampe. 2005. *An Annotated Catalog of the Native and Naturalized Flora of Arizona.* Flagstaff: The Arizona Ethnobotanical Research Association.

Integrated Taxonomic Information System. On-line database. http://www.itis.usda.gov (Date retrieved in 2005 and 2006).

Johnson, R. R. 1991. Historic changes in vegetation along the Colorado River. In *Colorado River Ecology and Dam Management*, edited by G. R. Marzolf, 178–206. Washington, D.C.: National Academy Press.

Kartesz, J., and C. Meacham. 2003. *A Synthesis of the North American Flora.* CD-ROM. Chapel Hill: North Carolina Botanical Garden.

Kearney, T. H., and R. H. Peables. 1960. *Arizona Flora.* Berkeley: University of California Press.

Kindscher, K. 1992. *Medicinal Wild Plants of the Prairie.* Lawrence: University of Kansas Press.

Kirkpatrick, Z. M. 1992. *Wildflowers of the Western Plains.* Austin: University of Texas Press.

Ku, M. S. B., J. Wu, Z. Dai, R. A. Scott, C. Chu and G. E. Edwards. 1991. Photosynthetic and photorespiratory characteristics of *Flaveria* species. *Plant Physiology* 96:518-28.

Leadbetter, R. Apples of the Hesperides. http://www.pantheon.org/articles/a/apples.of.the.hesperides.html.

Lexico Publishing Group, LLC. 2006. www.dictionary.com.

Lowe, C. H. 1964. *Arizona's Natural Environment: Landscapes and Habitats.* Tucson: University of Arizona Press.

Makarick, L. J. 1999. *Draft Exotic Plant Management Plan for Grand Canyon National Park.* Grand Canyon, Ariz.: National Park Service.

Marks, M., B. Lapin and J. Randall. 1993. *Element Stewardship Abstract for* Phragmites australis. The Nature Conservancy. http://tncweeds.ucdavis.edu/esadocs/documnts/phraaus.rtf.

Mayes, V. O., and B. B. Lacy. 1989. *Nanise': A Navajo Herbal.* Tsaile, Ariz: Navajo Community College Press.

McDougall, W. B. 1973. *Seed Plants of Northern Arizona.* Flagstaff: Museum of Northern Arizona.

———. 1964. *Grand Canyon Wild Flowers.* Flagstaff: Museum of Northern Arizona.

———. 1947. *Plants of Grand Canyon National Park.* Grand Canyon, Ariz.: Grand Canyon Natural History Association.

McGinnies, W. G. 1981. *Discovering the Desert: Legacy of the Carnegie Desert Botanical Laboratory.* Tucson: University of Arizona Press.

McIntyre, Anne. 1996. *Flower Power: Flower Remedies for Healing Body and Soul.* New York: Henry Holt and Co.

McKelvey, S. 1947. *Yuccas of the Southwestern United States.* Part Two. Jamaica Plain: Arnold Arboretum of Harvard University.

Minckley, W. L., and P. J. Unmack. 2000. Western springs: their faunas and threats to their existence. In *Freshwater Ecoregions of North America,* edited by R. A. Abell, D. M. Olson, E. Dinerstein, and P. T. Hurley, 52-53. Washington, D.C.: Island Press.

Moerman, D. E. 1998. *Native American Ethnobotany.* Portland, Oregon: Timber Press.

Molles, M. C. 2002. *Ecology: Concepts and Applications.* 2nd ed. New York: McGraw-Hill.

Montgomery, W. L., and P. E. Pollak. 2004. *The Unity and Diversity of Life II: Lives of Multicellular Organisms* (Laboratory Manual). Plymouth, Mich.: Hayden-McNeil Publishing, Inc.

Moore, M. 2003. *Medicinal Plants of the Mountain West.* Santa Fe: Museum of New Mexico Press.

Mozingo, Hugh. 1987. *Shrubs of the Great Basin: A Natural History.* Reno: University of Nevada Press.

Nabhan, G. P. 2004. *Cross-Pollinations: The Marriage of Science and Poetry.* Minneapolis, Minn.: Milkweed Editions Credo Series.

Nekola, J. C. 1999. Paleorefugia and neorefugia: The influence of colonization on community pattern and process. *Ecology* 80:2459-73.

Patraw, P. M. 1979. *Preliminary Check-List of Plants of Grand Canyon National Park.* Grand Canyon, Ariz.: United States Department of the Interior, National Park Service, Grand Canyon National Park.

Phillips, Arthur Morton. 1979. *Grand Canyon Wildflowers.* Grand Canyon, Ariz.: Grand Canyon Natural History Association.

Phillips, B. G., A. M. Phillips III, and M. A. Schmidt Bernzott. 1987. *Annotated Checklist of Vascular Plants of Grand Canyon National Park*. Grand Canyon, Ariz.: Grand Canyon Natural History Association.

Phillips, S. J., and P. W. Comus, eds. 2000. *A Natural History of the Sonoran Desert*. Tucson, Arizona: Sonora Desert Museum Press.

Pinkava, D. 1999. Cactaceae. Part Three. *Cylindropuntia. Journal of the Arizona-Nevada Academy of Science* 32(1):32-47.

Press, F., and R. Siever. 2001. *Understanding Earth*. 3rd ed. New York: W. H. Freeman and Company.

Proctor, M., P. Yeo, and A. Lack. 1996. *The Natural History of Pollination*. Portland, Oregon: Timber Press.

Redmond, C. L. 1999. *Human Impact on Ancient Environments*, Tucson: University of Arizona Press.

Red Scales in the Sunset. 2003. http://www.botgard.ucla.edu/html/botanytextbooks/ economicbotany/Cochineal.

Ricketts, H. W. 1971. *Wildflowers of the United States-Southwest Region*. Vol.2. New York: McGraw-Hill.

Russell, S. A. 2002. *Anatomy of a Rose: Exploring the Secret Life of Flowers*. New York: Perseus Publishing

Salzer, M. W., V. A. S. McCord, L. E. Stevens, and R. H. Webb. 1996. The dendrochronology of *Celtis reticulata* in the Grand Canyon: assessing the impact of regulated river flow on tree growth. In *Tree Rings, Environment, and Humanity, Proceedings of the International Conference, Tucson, AZ, May 17–21, 1994*, edited by J. S. Dean, D. M. Meko, and T. W. Swetnam, 273-81. Tucson, Ariz.: Radiocarbon.

Sanderson, M. J., and Wojciechowsk, M. F. 1996. Diversification rates in a temperate legume clade: Are there so many species of *Astragalus* (Fabaceae)? *American Journal of Botany* 83:1488-1502.

San Juan School District. 2003. *Fraxinus anomala*. http://dine.sanjuan.k12.ut.us/heritage/ land/plants/pj/single_ash.html (Accessed May 9, 2003).

Spellenberg, R. 2003. *Sonoran Desert Wildflowers*. Guilford, Conn.: The Globe Pequot Press.

———. 1979. *Audubon Society Field Guide to North American Wildflowers-Western Region*. New York: Alfred A. Knopf.

Stevens, L. E. 1983. *The Colorado River in Grand Canyon: A Guide*. 5th ed. Flagstaff, Ariz.: Red Lake Books.

Stevens, L. E., and R. A. Bailowitz. Distribution of *Brechmorhoga* clubskimmers (Odonata: Libellulidae) in the Grand Canyon region. *Western North American Naturalist*. In press.

Stevens, L. E., and R. L. Huber. Biogeography of Tiger Beetles (Cicindelidae) in the Grand Canyon Ecoregion. Arizona and Utah. *Cicindela* 35:41-64.

Stevens, L. E., and V. J. Meretsky, eds. *Every Last Drop: The Ecology and Conservation of North American Springs*. Tucson: University of Arizona Press. In press.

Stockert, J. W., and J. W. Stockert. 1967. *Common Wildflowers of the Grand Canyon*. Salt Lake City, Utah: Wheelwright Press.

Taylor, J. L. 2000. *Populus fremontii*. In *Fire Effects Information System*, U.S. Department of Agriculture, Forest Service, Rocky Mountain Research Station, Fire Sciences Laboratory (Producer). http://www.fs.fed.us/database/feis/ (accessed December 11, 2003).

Taylor, T. E., and K. L. Taylor. 1992. *Checklist of the Selected Plants of the Grand Canyon Area*. Grand Canyon, Ariz.: Grand Canyon Natural History Association.

Trimble, Stephen. 1999. *Sagebrush Ocean: A Natural History of the Great Basin*. Reno: University of Nevada Press.

Turner, R. M., J. E. Bowers, and T. L. Burgess. 1995. *Sonoran Desert Plants: an Ecological Atlas*. Tucson: University of Arizona Press.

Turner, R. M. and M. M. Karpiscak. 1980. *Recent Vegetation Changes along the Colorado River between Glen Canyon Dam and Lake Mead, Arizona*. U.S. Geological Survey Professional Paper 1132.

USDA, USFS. 2006. Fire Effects Information System [online datebase]. http://www.fs.fed.us/database/feis.

USDA, NRCS. 2006. *The PLANTS Database*, Version 3.5 (http://plants.usda.gov). Data compiled from various sources by Mark W. Skinner. National Plant Data Center, Baton Rouge, Louisiana 70874-4490 USA.

USGS Canyonlands Research Station. Biological soil crust website. http://www.soilcrusts.org.

Van Devender, T. 1987. Holocene vegetation and climate in the Puerto Blanco Mountains, southwestern Arizona. *Quaternary Research* 27:51-72.

Veilleux, C., and S. R. King. 2006. *An Introduction to Ethnobotany*. http://www.accessexcellence.org/RC/Ethnobotany/.

Warren, K. D. 1998. *Wild About Wildflowers*. Fort Collins, Colo.: Heel and Toe Publications.

Warren, P. L. 1982. *Vegetation of Grand Canyon National Park*. Tucson: University of Arizona Press.

Watahomigie, L. J., et al. 1982. *Hualapai Ethnobotany*. Peach Springs, Ariz.: Hualapai Bilingual Program, Peach Springs School District.

Webb, R. H. 1996. *Grand Canyon: A Century of Change: Rephotography of the 1889-1890 Stanton Expedition*. Tucson: University of Arizona Press.

Welsh, S. L., N. D. Atwood, S. Goodrich, and L. C. Higgins, eds. 1997. *A Utah Flora*. Provo, Utah: Brigham Young University.

Whitney, Stephen. 1982. *A Field Guide to the Grand Canyon*. New York: Quill Pubishers.

Whitson, T. D., L. C. Burrill, S. A. Dewey, D. W. Cudney, B. E. Nelson, R. D. Lee, and R. Parker. 1996. *Weeds of the West*. Newark, Calif.: Western Society of Weed Science in Cooperation with Western United States Land Grant Universities Cooperative Extension Services.

Wikimedia Foundation, Inc. 2006. Wikipedia: *The Free Encyclopedia*. http://en.wikipedia.org/wiki/Phenolic_compounds.

Wyman, L. C., and S. K. Harris. 1951. *The Ethnobotany of the Kayenta Navajo*. Albuquerque: University of New Mexico Press.

Zomlefer, Wendy B. 1994. *Guide to Flowering Plant Families*. Chapel Hill: University of North Carolina Press.

INDEX

Abronia nana, 118
    elliptica, 118–19
Acacia greggii, 13, 68–69
acacia, catclaw, 13, 68–69
Acanthoscelides species, 73
Aceraceae, 64
Acer negundo var. *arizonicum,* 64–65
    negundo ssp. *californicum,* 64
Achnatherum hymenoides, 17, 32–33
Acourtia wrightii, 186–87
Adenophyllum porophylloides, 220
Adiantum capillus-veneris, xvi, 27
Agavaceae, 78–83
Agave family, 78–83, 150
Agave species, xv, 16, 21, 78–79
    phillipsiana, 16, 17, 78–79
    utahensis var. *kaibabensis,* 78–79
    utahensis var. *utahensis,* 16, 78–79
alfalfa, 158–59
algae, 18
Alhagi camelorum, 156
    maurorum, 156–57
alien species, 23–24
alkali sacaton, 50–51
Allioni, Carlo Ludovico, 120
Allionia incarnata, 120–21
Allium bigelovii, 128–29
almond, desert, 146
Aloysia wrightii, 224–25
amaranth, pale, 84–85
amaranth family, 84
Amaranthaceae, 84
Amaranthus albus, 84–85
    cruentus, 84
ambersnail, Kanab, 142
Ambrosia acanthocarpa, 186
    dumosa, 11, 186–87
Ammospermophilus species, 152
Amsinckia menziesii var. *intermedia,* 88–89

Anacardiaceae, 184
Andropogon barbinodis, 36
    gerardii, 32–33
    glomeratus, 32–33
Anemone tuberosa, 96–97
annuals, 14
ants 216
Anulocaulis leiosolenus var. *leisolenus,* 120–21
Apache plume, 13, 172–73
Apiaceae, 110
Apocynaceae, 112
Apocynum cannabinum, 17, 112–13
Aquilegia chrysantha, 27, 96–97
    desertorum, 97
    micrantha, 96
Archilochus alexandri, 146
Argemone arizonica, 168
    munita, 168–69
Aristida adscensionis, 34–35
    arizonica, 34–35
    purpurea, 34
arrowleaf, white-flowered, 212
arrow-weed, 13, 26, 214–15
Artemisia frigida, 188
    ludoviciana, 188–89
    tridentata, 188
Asclepiadaceae, 134
Asclepias latifolia, 134–35
    subverticillata, 134–35
ash, single-leaf, 12, 66
    velvet, 66
aster, gray, 204–5
    lacy tansy, 210
    Mojave, xvi, 224–25
    spiny, 196–97
Aster glaucodes, 204
    spinosus, 196–97
Asteraceae, 186–225
Asterocampa species, 64

Astragalus lentiginosus, 156–57
    nuttallianus, 156–57
    praelongus, 156–57
Astrolepis cochisensis ssp. *cochisensis*, 28–29
athel, 74
Atriplex canescens, 24, 124–25
    confertifolia, 124–25
    obovata, 124

Baccharis species, 13, 188–91
    emoryi, 190–91
    glutinosa, 190
    salicifolia, 190–91
    sarothroides, 189, 190–91
    sergiloides, 190–91
baccharis, desert, 190–91
    seep-willow, 190–91
Bailey, Jacob Whitman, 192
Baileya multiradiata, 192–93
barberry family, 84
barberry, Fremont, 84
    red, 84–85
barrel cactus, California, 11, 22, 104–5
    many-headed, 100–101
bats, 78
beardgrass, bushy, 32–33
    silver, 36–37
beargrass, 150
beavertail cactus, Grand Canyon, 12, 106–7
Bebbia juncea var. *aspera*, 192–93
bedstraw, desert, 130–31
bees, 22, 78, 88, 112, 136, 158, 192, 202, 224
    bumble, 162, 198
    carpenter, 57, 152, 162
    digger, 170
    honey, 22
    leafcutter, 70
beetles, 198
    bruchid, 73
    chrysomelid, 88
    leaf, 112
    tiger, 20
bellflower family, 86
Berberidaceae, 84
Berberis haematocarpa, 84
bermuda grass, 40–41

Bert's Canyon, 168
Betulaceae, 58
Bigelow, John Milton, 128
Bignoniaceae, 57
bignonia family, 57
biological soil crusts, 19
birch family, 58
birchleaf buckthorn, 60–61
bittersweet family, 58, 88
blackbrush, 170–71
blackbrush community, 114, 170
blazing star, 182–83
bluedicks, 128–29
bluestem, big, 32–33
    cane, 36–37
    little, 36–37
Bombus species, 162
borage family, 88–90
Boraginaceae, 88–90
Bothriochloa barbinodis, 36–37
    saccharoides, 36-37
Boucher Canyon, 28
Bouteloua aristidoides, 36–37
    barbata, 36-37
    curtipendula, 36–37
boxelder, Arizona, 64–65
Brassica tournefortii, 138–39
Brassicaceae, 138–45
Brechmorhoga pertinax, 20
Brickell, John, 194
brickellbush, California, 194
    Coulter's, 194
    long-leaf, 194–95
    spiny, 194–95
Brickellia atractyloides, 194–95
    californica, 194–95
    coulteri, 194
    longifolia, 194–95
Bright Angel Creek, 20, 42, 132, 136
Bright Angel Trail, 28, 84, 226
brittlebush, 12, 202–3
    resin, 202–3
brome, red, 23, 38–39
    ripgut, 23, 38–39
Bromus species, 23, 38–39
    diandrus, 38–39

rigidus var. *gussonei*, 38
    rubens, 38, 39
    tectorum, 38–39
brooklime, American, 178
broom, desert, 188–91
broomrape, desert, 92–93
broomrape family, 92
brownfoot, 186–87
Buck Farm Canyon, 97, 136, 168, 198
buckhorn cholla, 100
buckthorn, birchleaf, 60–61
buckthorn family, 60, 92
buckwheat family, 94
buckwheat, 94–95
    Heermann's wild, 94–95
    shrubby wild, 94–95
buffaloberry, roundleaf, 13, 154–55
    silver, 154
buffalo fur, 186–87
bulrush, 56
bursage, white, 11, 186–87
burweed, annual, 186
buttercup family, 96–98
butterflies, 64, 88, 112, 134, 192, 198, 202, 224

Cactaceae, 98–109
cactus, 14, 98–99. *See* barrel cactus; cholla; claret-cup; hedgehog; prickly-pear
cactus family, 98–109
calico, bristly, 164–65
    Schott's, 164
Calochortus flexuosus, 130–31
    nuttallii, 130
caltrop family, 110
camelthorn, 156
Camissonia multijuga, 116–17
    specuicola, 116
Campanulaceae, 86
Canotia holacantha, 58–59
Carbon Canyon, 208
Carex species, 54–55
    aquatilis, 54–55
Cardenas Creek, 136
Cardenas Marsh, 76
cardinal flower, 86–87

carrot family, 110
Cassia covesii, 162
Castilleja applegatei ssp. *martinii*, 174–75
    linariifolia, 174
Castillejo, Domingo, 174
catclaw acacia, 68–69
caterpillars, 64, 112, 144–45, 224
caterpillar weed, 226
cattail, broad-leaved, 30–31
    southern, 30–31
cattail family, 30
cedar, pygmy-, 214–15
Celastraceae, 58, 88
Celtis laevigata var. *reticulata*, 13, 64–65
    reticulata, 64
Centaurium arizonicum, 122–23
    calycosum var. *arizonicum*, 122
centaury, Arizona, 122–23
Centris species, 170
century plant, Grand Canyon, 16, 17, 78–79
    Kaibab, 78–79
    Utah, xv, 16, 78–79
Cercis occidentalis var. *orbiculata*, 70
    orbiculata, 13, 70–71
Cercocarpus montanus, 172
Chaenactis stevioides, 196–97
Chamisso, A. L. Von, 116
Chamaesyce fendleri, 180–81
cheatgrass, 38–39
checkerspot, desert, 224
Cheilanthes feei, 28–29
Chenopodiaceae, 124–27
chicory, desert, 218–19
Chihuahuan Desert, 10, 11
Chilopsis linearis, 57
Chloracantha spinosa, 196–97
Chlosyne neumoegeni, 224
cholla, buckhorn, 100
    Peach Springs Canyon, 98
    teddy-bear, 98–99
    whipple, 100–101
Chrysochus auratus, 112
chrysomelid beetles, 88
chuckwalla's delight, 192–93
chuparosa, 192
Cicindela hemorrhagica arionae, 20

Cirsium arizonicum var. *bipinnatifita*, 198
    mohavense, 198
    neomexicanum, 198–99
    rydbergii, 198–99
citrus family, 60, 112
Cladium californicum, 25, 54–55
Cladophora glomerata, 18
claret-cup cactus, 102–3
Clear Creek, 42, 122, 132, 156, 198
cliff-brake, spiny, 28–29
cliffrose, 172
Clover, Elzada, iii, 102, 106
cochineal-scale insect, 106
cocklebur, 222-23
coldenia, shrubby, 90–91
Coldenia hispidissima var. *latior*, 90
Coleogyne ramosissima, 170–71
Collom, Rose, iii
Colorado Plateau, 11
Colorado River: dam-managed flows of, 13, 18, 25; migration of plants up, 11; plants of pre-dam high-water line, 12, 13, 68, 70, 72, 154, 172
columbine, alcove, 96
    golden, 27, 96–97
Compositae, 186–225
Conyza canadensis, 200–201
Cortaderia species, 48–49
cotton-top cactus, 100
cottonwood, Fremont, xvi, 24, 74–75
Cove Canyon, 150
coyote willow, 26
creeper, Virginia, 128–29
creosotebush, 11, 14, 110-11
Cruciferae, 138–45
crucifixion-thorn, 58–59
cryptantha, 90–91
    woody, 90
Cryptantha species, 90–91
    racemosa, 90
cudweed, 216–17
    Wright, 216
Cupressaceae, 62
Cutter, Lois Jotter, iii, 13, 102, 106
Cylindropuntia species, 46, 98
    abyssi, 98

    acanthocarpa, 100-101
    bigelovii var. *bigelovii*, 98–99
    whipplei, 100–101
Cymopterus purpurascens, 110–111
Cynodon dactylon, 40–41
Cyperaceae, 54–55
Cyperus species, 54–55
    erythrorhizos, 55
cypress family, 62

Dactylopius coccus, 106
daisy, Emory rock, 212-13
    fleabane, 204–5
    Grand Canyon rock, 212-13
Dalea arborescens, 160
Danaus plexipus, 134
dandelion, common, 212
    desert-, 212-13
Dasyochloa pulchellum, 42–43
datura, sacred, xii, 22, 144-45
Datura meteloides, 144
    wrightii, xii, 22, 144–45
Deer Creek, 20, 27, 74, 92, 116, 148, 150, 172, 184
Delphinium parishii, 98–99
    scaposum, 98
Descourain, Francois, 138
Descurainia pinnata, 138–39
    sophia, 138
desert-dandelion, 212–13
desertscrub community, 9, 10
desert-willow, 57
devil's claw, 22
Diamond Creek, 76, 98
Diamond Creek Road, 57, 58
diatoms, 18
Dichelostemma pulchellum var. *pauciflorum*, 128–29
dicoria, desert, 200-201
Dicoria canescens, 200-201
Distichlis spicata, 40-41
dogbane, 17, 112–13
dogbane family, 112
dogweed, 220–21
    needleleaf, 220
domesticated plants, 16

dragonfly, masked clubskimmer, 20
draba, wedge-leaf, 140–41
Draba cuneifolia, 140–41
dropseed, giant, 50-51
    mesa, 50-51
    sand, 50-51
    spike, 50-51
dunebroom, common, 158–59
dyssodia, San Felipe, 220
Dyssodia pentachaeta var. *pentachaeta*, 220

Echinocactus polycephalus var. *polycephalus*, 101
    var. *xeranthemoides*, 100–101
Echinocereus coccineus, 102
    engelmannii, 102–3
    triglochidiatus, 102–3
ecosystems, relict, 20
Elaeagnaceae, 154
Elaeagnus angustifolia, 23, 154
Eleocharis species, 56
    parishii, 56
elm family, 64
Elodea species, 18
Elves Chasm, x, 28, 172
Emory, William Hemsley, 190, 212
Empidonax traillii extimus, 76
Engelmann, George, 102
Encelia farinosa, 12, 202–3
    resinifera, 202–3
endemic, 20, 25
Entzelt, Christopher, 202
ephedra, 114-15
    Cutler's, 114
    green, 114
    Mojave, 114
    Torrey, 114
ephedra family, 114
Ephedra species, 114–15
    cutleri, 114
    fasciculata, 114–15
    nevadensis, 114
    torreyana, 114–15
    viridis, 114
Ephedraceae, 114
Epipactis gigantea, 154–55
Equisetaceae, 26

Equisetum x ferrissii, 26-27
    hyemale, 26
    laevigatum, 26
Erianthus ravennae, 48
eriastrum, diffuse, 164–65
Eriastrum diffusum, 164–65
Ericameria nauseosa, 208–9
Erigeron lobatus, 204–5
Eriogonum corymbosum, 94
    deflexum, 94–95
    heermannii var. *subracemosum*, 94
    inflatum, 95
Erioneuron pulchellum, 42
Erodium cicutarium, 122–23
    texanum, 122–23
Eschscholzia minutiflora, 168–69
Eucnide urens, 182–83
Eucrypta micrantha, 228
eucrypta, small-flowered, 228
Euphorbia aaron-rossii, 180–81
    fendleri, 180
Euphorbiaceae, 180
Eurybia glauca, 204–5
Euthamia occidentalis, 206–7
Euura species, 76
evening-primrose, Hooker's, 116–17
    longstem, 116
    pale, 118–19
    tufted, 118–19
evening-primrose family, 116–19
exotic, 23–24

Fabaceae, 68–73, 156–63
Fagaceae, 66
Fallugi, Virgilio, 172
Fallugia paradoxa, 13, 172–73
Fendler, Augustus, 180
Fendler's sandmat, 180–81
fern, scaly cloak, 28–29
    maidenhair, xvi, 27
    slender lip, 28–29
Fern Glen Canyon, 172
ferns and fern allies, 26–29
Ferocactus acanthodes, 104
    cylindraceus, 11, 22, 104–5
fescue, six-weeks, 52–53

Festuca octoflora, 52
fetid-marigold, 220–21
fiddleneck, common, 88–90
fiesta flower, Arizona, 228
filaree, 122–23
fire, 23
firecracker, Eaton's, 178–79
fishhook cactus, Graham's, 104–5
Flaveria macdougallii, 20, 206–7
flaveria, Grand Canyon, 20, 206–7
fleabane daisy, 204–5
flixweed, 138
fluffgrass, 42–43
flycatcher, willow, 76
Fourquier, Pierre Éloi, 152
Fouquieria splendens, 11, 15, 152–53
Fouquieriaceae, 152
four o'clock, desert, 120–21
    trailing, 120–21
four o'clock family, 118–21
Frangula betulifolia ssp. *obovata*, 60–61
Franseria dumosa, 186
Fraxinus anomala, 12, 66-67
    pennsylvanica ssp. *velutina*, 66
    velutina, 66-67
Fremont, John Charles, 75
Funastrum cynanchoides ssp. *cynanchoides*, 134

Galium aparine, 130
    stellatum var. *eremicum*, 130–31
galleta, 46–47
    big, 46–47
Gambel oak, 66–67
gentian family, 122
Gentianaceae, 122
geologic time, 8
Geraniaceae, 122
geranium family, 122
gilia, eyed, 162–63
Gilia ophthalmoides, 162–63
Glen Canyon Dam, 12, 13, 18, 25, 30, 160, 214
Glen Canyon National Recreation Area, 138
globemallow, desert, 132
    gooseberryleaf, 132–33

glowweed, willow, 208–9
Gnaphalium chilense, 216
    wrightii, 216
goldenbush, 210
    Grand Canyon, 208
goldenrod, western, 206–7
goldenweed, spiny, 210–11
Goodding, Leslie Newton, 76
goosefoot family, 124–27
grama, sideoats, 36–37
    six-weeks, 36–37
    six-weeks needle, 36–37
Graminae, 32–53
Granite Park, 76
grape family, 128
grape, Arizona, 128–29
    canyon, 128–29
grass family, 32–53
grasslands, desert, 9
Gray, Asa, 186, 224
graythorn, 92–93
Great Basin Desert, 10, 11
Gregg, Josiah, 68
ground cherry, ivy-leaved, 148
    thick-leaved, 148–49
Gutierrezia microcephala, 208–9
    sarothrae, 208–9

hackberry, netleaf, 13, 64–65
hanging garden, x
Hance Trail, 146
Haplopappus salicinus, 208
    spinulosus, 210
Havasu Canyon, 21–22, 28, 57, 66, 84, 92, 114, 146, 156, 174, 198
Havasupai people, 16, 27, 101, 154
Havasupai Village (Supai), 16
hawk moth, 22, 116, 144, 198
Hedeoma drummondii, 136
    nanum, 136–37
    oblongifolium, 136
hedgehog, Engelmann, 102–3
Helianthus annuus, 222
helleborine orchid, 154–55
hemp, Indian, 112
Hermit Creek, 66

Hermit Trail, 146, 156
Hesperodoria salicina, 208–9
    scopulorum var. *scopulorum*, 208
Hesperoyucca newberryi, 80–81
Hilaria jamesii, 46
Hooker, Joseph Dalton, 116
hop hornbeam, western, 58–59
Hopi people, 46, 75, 84, 125, 126
hoptree, pale, 60–61
hornworm, 144–45
horsetail family, 26
horsetail, 26-27
horseweed, 200–201
Hualapai people, 76–79
hummingbirds, 21, 78, 146, 152, 198
Hydrophyllaceae, 226-28

Imperata brevifolia, 25, 42–43
Indian hemp, 112
Indian ricegrass, 17, 32–33
indigobush, Fremont, 160
    Mojave, 160–61
inkweed, 126
invertebrates, microscopic, 18
Isocoma acradenia, 210–11
invasives, 23-24
Ives, Joseph, 80

Jaeger, Edmund, 192
Janusia gracilis, 132–33
janusia, slender, 132–33
jimmyweed, 210
jimson weed, 144-45
Joshua-tree, 80–81
Judas tree, 70
Juncaceae, 52–53
Juncus articulatus, 52–53
    balticus, 52–53
    torreyi, 52–53
juniper, common, 62
    one-seed, 62–63
    Utah, 62
Juniperus species, 12, 62
    communis, 62
    monosperma, 62–63
    occidentalis var. *gymnocarpa*, 62
    osteosperma, 62

Kaibab Trail, 60. *See also* North Kaibab Trail
Kanab Canyon, 78, 84
Knowlton, Frank, 58
Krameria erecta, 170–71
    parvifolia, 170
krameria, 170–71
Krameriaceae, 170
Kwagunt Canyon, 66

Labiatae, 136
lac insects, 110
Lactuca serriola, 218
Lake Mead, 104, 138
Lake Powell, 160
Lamiaceae, 136
Langlois, Auguste Barthelemy, 164
Langloisia setosissima ssp. *setosissima*, 164–65
larkspur, barestem, 98
    desert, 98–99
Larrea divaricata, 110
    tridentata var. *tridentata*, 11, 14, 110–11
Lava Falls, 19
Lees Ferry, 24, 94, 126, 136, 138, 158
Leguminosae, 68–73, 156–63
lemonade-berry, 184
Lepidium fremontii, 140–41
    latifolium, 140–41
    montanum, 140
lettuce, devil's, 88
    wild, 218
    wire, 220–21
life zone, 9, 10
Liliaceae, 128–31
lily family, 128–31
lily, Mariposa, 130-31
    white-flowered sego, 130
linanthus, Bigelow's, 164–65
Linanthus bigelovii, 164-65
lippia, Wright's, 224-25
Lippia wrightii, 224-25
Little Colorado River, 10, 30, 100, 156
live oak, shrub, 66–67
Loasaceae, 182
L'Obel, Matthias de, 86
lobelia, scarlet, 86–87
Lobelia cardinalis, 86–87
locoweeds, 156

Loeseliastrum schottii, 164
London rocket, 138–39
Lycium andersonii, 146-47
    fremontii, 146
    pallidum, 146–47

Machaeranthera pinnatifida, 210-11
    tortifolia, 224
macrophytes, 18
madder family, 130
Mahonia fremontii, 84
    haematocarpa, 84–85
maidenhair family, 27–28
Malacothrix californica var. *glabrata*, 212
    glabrata, 212–13
    sonchoides, 212
mallow family, 132
Malpighia family, 132
Malpighiaceae, 132
Malvaceae, 132
Mammallaria grahamii var. *grahamii*, 104–5
    microcarpa, 104
    tetrancistra, 104
maple family, 64
Marble Canyon, 10, 63, 108, 180
marigold, desert, 192–93
marigold, fetid-, 220–21
Mariposa lily, 130–31
Matkatamiba Canyon, 60, 84, 136
Maurandella antirrhiniflora, 176–77
Maurandy, Catalina Pancracia, 176
Maurandya antirrhiniflora, 176
McDougall, Walter B., 206
Medicago sativa, 158–59
Mediterranean grasses, 52–53
Megachile species, 70
Melilotus alba, 158
    officinalis, 158–59
Mentha arvensis, 136–37
Mentzel, Christian, 182
Mentzelia pumila, 182–83
mesquite, western honey, 13, 14, 17, 72–73
milkvetch, small-flowered, 156–57
    specklepod, 156-57
    stinking, 156-57

milkweed, broadleaf, 134-35
    twining, 134–35
    western whorled, 134
milkweed family, 134
Mimulus cardinalis, 20, 27, 176–77
    guttatus, 176–77
mint, field, 136–37
mint family, 136
Mirabilis multiflora, 120-21
mistletoe, 73
mock pennyroyal, dwarf, 136–37
Mojave Desert, 10, 11
monkeyflower, cardinal, 20, 27, 176–77
    yellow, 176–77
Mormon tea, 114
Morton, Samuel George, 88
mortonia, Utah, 88–89
Mortonia scabrella, 88
    scabrella var. *utahensis*, 88
    utahensis, 88–89
moth, 95
    hawk, 116, 144
    schinia, 192, 224
    yucca, 21, 80, 82
mountain mahogany, 172
Muhlenberg, G. H. E., 44
Muhlenbergia asperifolia, 44–45
mustard, Sahara, 138–39
    tumble, 138
    yellow tansy, 138-39
mustard family, 138–45

Nankoweap Canyon, 46, 66, 126
Nankoweap Trail, 146
Nasturtium officinale, 142
Navajo people, 46, 154
Nemacladus glanduliferus ssp. *arizonica*, 86–87
nest-straw, desert, 216
Newberry, John Strong, 80
Newberry's yucca, 80–81
Nicot, Jean, 148
Nicotiana obtusifolia var. *obtusifolia*, 17, 148–49
    trigonophylla, 148
nightshade family, 144–51

nightshade, American, 150–51
    silverleaf, 150–51
Nolina family, 150
Nolina microcarpa, 150–51
Nolinaceae, 150
nopal, 107
North Canyon, 60, 168
North Kaibab Trail, 168, 174, 184, 226
Notholaena cochisensis, 28
Nyctaginaceae, 118–21

oak, Gamble, 66–67
    shrub live, 66–67
oak family, 66
Oarisma garita, 192
Ochlodes yuma, 192
ocotillo, 11, 15, 152–53
ocotillo family, 152
Oenothera caespitosa, 118–19
    elata ssp. *hookeri*, 116–17
    hookeri, 116
    longissima, 116
    pallida, 118–19
Oleaceae, 66
oleaster family, 154
olive, Russian, 23, 154
olive family, 66
Onagraceae, 116–19
onion, Bigelow, 128–29
Opuntia species, 23, 106–9
    basilaris var. *longiareolata*, 12, 106–7
    bigelovii, 98
    chlorotica, 108
    engelmannii var. *engelmannii*, 107
    phaeacantha, 108–9
    polyacantha var. *erinacea*, 108–9
    polyacantha var. *hystricina*, 108–9
    polyacantha var. *nicholii*, 108
    whipplei, 98
orchid, helleborine, 154–55
Orchidaceae, 154
orchid family, 154
Orobanchaceae, 92
Orobanche cooperi, 92–93
Oryzopsis hymenoides, 32
Ostrya knowltonii, 58–59

Oxyloma haydeni kanabensis, 142

pack rat middens, 12, 88, 98, 101
paintbrush, common, 174–75
    long-leaved, 174
Palisades Creek, 126
Palmer, Ernest Jesse, 178
pampus grass, 48–49
Panicum capillare, 44–45
Papaveraceae, 168
Papilio species, 112
Parryella filifolia, 158–59
parsley, wide-wing spring-, 110–11
Parthenocissus vitacea, 128–29
Peach Springs Canyon, 98
pea family, 68–73, 156–63
Pellaea longimucronata, 28
    truncata, 28–29
Penstemon eatonii, 178–79
    palmeri, 178–79
pennyroyal, dwarf mock, 136–37
penstemon, Palmer's, 178–79
pepperweed, broadleaved, 140–41
    desert, 140–41
    mountain, 140–41
Perezia wrightii, 186
Perityle congesta, 212–13
    emoryi, 212–13
perennials, 14
Petrophyton caespitosum, 172–73
Peucephyllum schottii, 214–15
Phacelia crenulata, 226–27
    filiformis, 226–27
    glechomifolia, 226–27
Phantom Ranch, 150, 198
phlox family, 162–65
Pholistoma auritum ssp. *arizonicum*, 228
Phoradendron californicum, 73
phragmites, xvi, 44–45
Phragmites australis, xvi, 44–45
    communis, 44
phreatophytes, 14
Physalis crassifolia, 148–49
    hederifolia, 148
    ixocarpa, 148
pigweed, tumble, 84–85

pincushion, corkseed, 104
    desert, 196–97
pincushion cactus, 104
pine, piñon, 62
Pinus edulis, 62
Piptatherum miliaceum, 32–33
Plantaginaceae, 166
Plantago lanceolata, 166–67
    major, 166–67
    ovata, 166–67
    patagonica, 166–67
    purshii, 166
plantain, blonde, 166–67
    common, 166–67
    lanceleaf, 166–67
    woolly, 166–67
plantain family, 166
Pleuraphis jamesii, 46–47
    rigida, 46–47
Pleurocoronis pluriseta, 212
Pluchea sericea, 13, 26, 214–15
Poaceae, 32–53
poison-ivy, 184–85
Polemoniaceae, 162–65
pollinators, 21–22. See also specific pollinator names
Polygonaceae, 94–95
polypogon, ditch, 46–47
Polypogon interruptus, 46–47
    monspeliensis, 46–47
    viridis, 46–47
poppy, little gold, 168
    sore-eye, 132
poppy family, 168
Populus fremontii, xvi, 24, 74–75
poreleaf, 216–17
poreweed, 216–17
Porophyllum gracile, 216–17
Powell, John Wesley, 64, 72
prickly-pear, 23, 106–9
    brown-spined, 108–9
    Engelmann, 107
    grizzly-bear, 108–9
    Navajo Bridge, 108
    pancake, 108–9
prickly-poppy, 168–69
    Roaring Springs, 168

primrose, cave, 168–69
Primrose family, 168
Primula hunnewellii, 168
    specuicola, 168–69
Primulaceae, 168
prince's-plume, 17, 142–43
Proboscidea parviflora, 22
Prosopis glandulosa var. *torreyana*, 13, 14, 17, 72–73
    juliflora, 72
Prunus fasciculata, 146
Pseudognaphalium canescens ssp. *canescens*, 216
    stramineum, 216–17
Psoralea juncea, 160–61
Psoralidium junceum, 25, 160–61
Psorothamnus arborescens, 160–61
    fremontii, 160–61
Ptelea trifoliata ssp. *pallida*, 60–61
Pteridaceae, 27–28
Purshia stansburiana, 172
pygmy-cedar, 214–15

Quercus gambellii, 66–67
    turbinella, 66-67

rabbitbrush, 208–9
rabbitfoot grass, 46–47
Rafinesque-Schmaltz, C. S., 218
Rafinesquia neomexicana, 218–19
ragweed, 186, 222
Ranunculaceae, 96–98
ratany, littleleaf, 170
ratany family, 170
ravenna grass, 48–49
redbud, California, 70–71
    western, 13, 70–71
reed, common, 44-45
Rhamnaceae, 60, 92
Rhamnus betulifolia var. *obovata*, 60
Rhus glabra, 184
    trilobata var. *simplicifolia*, 184–85
    trilobata var. *trilobata*, 184
Rider Canyon, 168
ricegrass, Indian, 32–33
ringstem, 120–21
riparian community, 9, 13, 20, 23
Roaring Springs, 20, 174

rock daisy, Emory, 212–13
    Grand Canyon, 212-13
rocket, London, 138–39
rock nettle, desert, 182–83
rockmat, 172–73
rockspirea, mat, 172
Rorippa nasturtium-aquaticum, 20, 142–43
Rosa arizonica, 174
    stellata, 174
    woodsii var. *ultramontana*, 174–75
Rosaceae, 170–75
rose family, 170–75
rose, Arizona, 174–75
    desert, 174
    wild, 174–75
Ross, Aaron B., 180
Ross's spurge, 180–81
Rubiaceae, 130
rush family, 52–53
rush, jointed, 52–53
    scouring, 26
    Torrey's, 52–53
    wire, 52–53
rushlike scurf-pea, 25, 160–61
Rutaceae, 60, 112
Rydberg, Per Axel, 184

sacaton, alkali, 50–51
Saccharum ravennae, 48–49
saddlebush, Rio Grande, 88
Saddle Canyon, 23, 60, 84, 116, 186, 198, 208
sage, Davidson's, 136–37
    fringed, 188
    purple, 136–37
sagebrush, big, 188
Salicaceae, 74–77
Salix species, 13, 24
    exigua, 26, 76–77
    gooddingii, 76–77
Salsola iberica, 126
    tragus, 126–27
saltbush, four-wing, 24, 124–25
    New Mexico, 124
saltcedar, 74–75
saltgrass, inland, 40–41

Salvia davidsonii, 136
    dorrii ssp. *dorrii*, 136–37
sandmat, Fendler's, 180–81
sandpaper-bush, 88–89
sand-verbena, 118–19
    dwarf, 118
Sarcostemma cynanchoides ssp. *cynanchoides*, 134-35
    hirtellum, 134
satintail, 25, 42–43
saucers, yellow, 212
sawfly, 76
sawgrass, 25, 54–55
Schinia miniana, 192
Schinia species, 193
Schismus species, 52–53
Schizachyrium scoparium, 36–37
Schoenoplectus species, 56
Schott, Heinrich Wilhelm, 214
Schott's calico, 164
Scirpus species, 56
scorpionweed, Grand Canyon, 226
    notch-leaf, 226-27
    pennyroyal-leaf, 226
scratchgrass, 44-45
Scrophulariaceae, 174–79
scurf-pea, rushlike, 25, 160–61
sedge family, 54–56
sedge, flat, 54–55
    leafy, 54–55
seepweed, desert, 126–27
    woody, 126
seep-willow, 190–91
seep-willow, Emory's, 190–91
Selasphorus platycercus, 21
Senna covesii, 162–63
senna, desert, 162–63
shadscale, 124–25
Shepherdia argentea, 144
    rotundifolia, 13, 154–55
Shepherd, John, 154
sideoats grama, 36–37
Sisymbrium altissimum, 138
    irio, 138–39
skeleton weed, 94–95
skipper, Yuma, 192

skipperling, western, 192
skunkbush, 184–85
smilograss, 32–33
snakeweed, broom, 208–9
    threadleaf, 208–9
snapdragon family, 174–79
snapdragon, twining, 176–77
soaptree yucca, 82–83
soil crust, 19
Solanaceae, 144–51
Solanum americanum, 150–51
    elaeagnifolium, 150–51
Solidago occidentalis, 206
Sonchus asper, 218–19
    oleraceus, 218–19
Sonoran Desert, 10, 11
sow-thistle, common, 218–19
    spiny, 218–19
speedwell, water, 178–79
Sphaeralcea ambigua, 132
    grossulariifolia, 132-33
Sphingidae, 144
spikerush, 56
Sporobolus airoides, 50–51
    contractus, 50–51
    cryptandrus, 50–51
    flexuosus, 50–51
    giganteus, 50–51
spring-parsley, wide-wing, 110-11
springs, 20
spurge family, 180
spurge, Marble Canyon, 180–81
    Ross's, 180–81
squirrel, antelope ground, 152
Stanley, Edward Smith, 142
Stanleya pinnata, 17, 142-43
Stanton, Robert Brewster, 19
Stanton's Cave, 28
Stephanomeria exigua, 221
    pauciflora, 220–21
    tenuifolia, 220–21
stickleaf, 182–83
stickleaf family, 182
sticky willy, 130
Stone Creek, 42, 74, 80, 82, 116, 136, 148, 184

storksbill, 122-23
    Texas, 122-23
Streptanthella longirostris, 144-45
Stylocline micropoides, 216
Suaeda moquinii, 126–27
    suffrutescens, 126
    torreyana, 126
sumac, smooth, 184–85
sumac family, 184
suncup, 116–17
    Kaibab, 116
sunflower family, 186–225
Supai, 16
swallowtail, 112
sweetclover, white, 158–59
    yellow, 158–59

Tachardiella larreae, 110
Tamaricaceae, 74
tamarisk, 13, 24, 26, 74–75
tamarisk family, 74
Tamarix aphylla, 74
    chinensis, 74–75
    pentandra, 74–75
    ramosissima, 13, 24, 26, 74–75
tansy mustard, yellow, 138–39
Tapeats Creek, 74, 136
Taraxacum officinale, 212
tea, mormon, 114–15
teddy-bear cholla, 98–99
Tegeticula species, 21, 82
    maculata, 80
Tessaria sericea, 214
Thamnosma montana, 112–13
Thelypodium integrifolium ssp. *longicarpum*, 142–43
thelypody, 142–43
thistle, Cainville, 198
    Mojave, 198
    New Mexico, 198–99
    Russian, 126
    Rydberg, 198–99
threadleaf, glandular, 86–87
three-awn, Arizona, 34–35
three-awn, purple, 34
three-awn, six-weeks, 34–35

Three Springs, 122
Thunder River, 136, 198
Thymophylla acerosa, 220
    pentachaeta var. *pentachaeta*, 220–21
tiquilia, matted, 90–91
Tiquilia canescens, 90–91
Tiquilia latior, 90–91
tobacco, desert, 17, 148—49
tomatillo, 148
Tonto Plateau, 106, 108, 110, 114, 128, 170
Toroweap Point, 224
Toxicodendron rydbergii, 184–85
Trixis californica, 222–23
trixis, 222-23
trumpet, desert, 94–95
tule, 56
tumbleweed, 126-27
turpentine-broom, 112-13
twin bugs, desert, 200-201
twist flower, long-beaked, 144-45
209 Mile Wash, 57
Typhaceae, 30
Typha domingensis, 30-31
    latifolia, 30

Ulmaceae, 64
Umbelliferae, 110

Vasey's Paradise, 30, 142, 184
Velcro plant, 182
verbena, lemon, 224-25
Verbenaceae, 224
veronica, 178-79
Veronica americana, 178
    anagallis-aquatica, 178–79
vervain family, 224
Vitaceae, 128
Vitis arizonica, 128–29
Vulpia octoflora, 52–53

Waltenberg Canyon, 46
wasps, 88

waterbent grass, 46–47
watercress, 20, 142-43
waterleaf family, 226–28
water-sage, 188–89
waterweed, 18
Whipple, Amiel Weeks, 100
Whitmore Trail, 170
willow family, 74–76
willow, coyote, 13, 24, 76–77
    desert-, 57
    Goodding's, 76–77
    sand-bar, 76–77
windflower, desert, 96–97
wire lettuce, 220–21
witchgrass, 44–45
wolfberry, Anderson's, 146–47
    Fremont, 146
    pale, 146–47
Woods, Joseph, 174
wormwood, Louisiana, 188–89
Wright, Charles, 186, 224

Xanthium strumarium, 222-23
xerophytes, 14
Xylocopa species, 57, 152, 162
Xylorhiza tortifolia, xvi, 224–25

yerba del venado, 216
Yucca species, 17, 21, 80
    baccata, 17, 82–83
    brevifolia, 80
    elata, 82–83
    whipplei, 80
yucca, banana, 17, 82–83
    narrow-leaved, 82
    Newberry's, 80–81
    soaptree, 82–83
    whipple, 80–81

Ziziphus obtusifolia var. *canescens*, 92–93
Zygophyllaceae, 110

ABOUT THE CONTRIBUTING WRITERS

An ecologist and educator on the Colorado Plateau for the past twenty years, **Emma Benenati** is also the Research Coordinator for Grand Canyon National Park.

Dan Hall has a maniacal thirst for killing invasive plants on the Colorado River, where he has been working as a boatman since 1988.

Marisa Howe studied the riparian vegetation along the banks of the Colorado River in the Grand Canyon for her graduate research.

A southwestern ethnobiologist and local-foods advocate for more than thirty years, **Gary Paul Nabhan** is also the author of over twenty books, many of which have won awards.

Art Phillips, a botanist based in Flagstaff, has studied the vegetation and flora of the Grand Canyon and the greater Southwest for more than thirty-five years.

Fred Phillips has been doing riparian restoration on the Colorado River for more than thirteen years, and he would like to revegetate the entire river corridor.

Joe Shannon, an aquatic ecologist and science educator, has worked and played in the canyons of the Colorado Plateau for the past twenty years.

Larry Stevens is the Curator of Ecology and Conservation at the Museum of Northern Arizona in Flagstaff, and he studies Grand Canyon's ecosystems and biota.

For the past fifteen years, as an archaeologist for the Hopi Tribe, **Michael Yeatts** has been involved in research and preservation of Hopi culture in the Grand Canyon.

Ann Zwinger writes extensively about the natural history of the Southwest. She has authored over ten books, including *Downcanyon: A Naturalist Explores the Colorado River through Grand Canyon* and *The Nearsighted Naturalist*. Ann currently lives in Colorado Springs, Colorado.

We encourage you to patronize your local bookstore. Most stores will order any title they do not stock. You may also order directly from Mountain Press, using the order form provided below or by calling our toll-free, 24-hour number and using your VISA, MasterCard, Discover or American Express.

Some other Natural History and Southwest titles of interest:

____Awesome Ospreys Fishing Birds of the World	$12.00
____Desert Wildflowers of North America	$24.00
____Edible and Medicinal Plants of the West	$21.00
____Finding Fault in California An Earthquake Tourist's Guide	$18.00
____Fire in the Sierra Nevada Forests	$20.00
____From Earth to Herbalist An Earth-Conscious Guide to Medicinal Plants	$21.00
____Geology Underfoot in Central Nevada	$16.00
____Geology Underfoot in Death Valley and Owens Valley	$18.00
____Geology Underfoot in Southern California	$14.00
____Geology Underfoot in Southern Utah	$18.00
____Introduction to Southern California Butterflies	$22.00
____Nature's Yucky Gross Stuff That Helps Nature Work	$10.00
____Plants of the Lewis & Clark Expedition	$20.00
____Roadside Geology of Arizona	$18.00
____Roadside Geology of Colorado, Second Edition	$20.00
____Roadside History of Arizona, Second Edition	$20.00
____Roadside History of Colorado	$20.
____Roadside History of Utah	$18.00
____Roadside Plants of Southern California	$15.00
____Rock Art Savvy A Responsible Visitor's Guide to Pubic Sites of the Southwest	$16.00
____Sagebrush Country A Wildflower Sanctuary	$14.00
____Sierra Nevada Wildflowers	$16.00
____The Southwest Inside Out An Illustrated Guide to the Land and Its History	$24.95
____Watchable Birds of the Great Basin	$16.00
____Watchable Birds of the Southwest	$14.00
____Weather Extremes of the West	$24.00
____Wild Berries of the West	$16.00

Please include $3.00 per order to cover shipping and handling.

Send the books marked above. I enclose $_____

Name_____

Address_____

City_____State_____Zip_____

☐ Payment enclosed (check or money order in U.S. funds)

Bill my: ☐ VISA   ☐ MasterCard   ☐ Discover   ☐ American Express

Card No._____Exp. Date:_____

Signature _____

MOUNTAIN PRESS PUBLISHING COMPANY

P.O. Box 2399 • Missoula, MT 59806
Order Toll Free 1-800-234-5308 • Have your credit card ready.
e-mail: info@mtnpress.com • website: www.mountain-press.com

ABOUT THE AUTHORS

Kristin Huisinga combines her master's degree in botany with a passion for community-based projects. She works with native people in the Southwest to incorporate indigenous perspectives about the land into management plans and educational programs. For pleasure, Kristin rows boats, tends to her high-elevation garden, walks to the top of steep hills, and sends at least one letter a week.

Lori Makarick grew up in New Jersey but began a steady, determined migration west as soon as she could drive. She has spent the majority of her adult life living and working in the Grand Canyon area and is committed to protecting its diverse natural resources. Lori consistently strives to find more time for travel and will one day attain a healthy balance between work and play.

Kate Watters first stumbled on the Grand Canyon's plant world while working on a trail crew there almost ten years ago where she quickly became transfixed by desert plants. Although botany and restoration ecology are her abiding passions, she occupies her idle moments with a variety of artistic endeavors. She lives in Flagstaff with her husband and cat.

Kristin Huisinga, Lori Makarick, and Kate Watters
—Photograph © 2006 by Raechel Running RMRfotoart.com